Quantum Computing and Communications

Edited by Yongli Zhao

Published in London, United Kingdom

IntechOpen

Supporting open minds since 2005

Quantum Computing and Communications
http://dx.doi.org/10.5772/intechopen.90976
Edited by Yongli Zhao

Contributors
Rene Steijl, Ismail Gassoumi, Lamjed Touil, Abdelatif Mtibaa, Graciana Puentes, Surya Teja Marella, Hemanth Sai Kumar Parisa, Yongli Zhao, Qingcheng Zhu, Yazi Wang, Lu Lu, Xiaosong Yu, Jie Zhang, Yuan Cao

First published in London, United Kingdom, 2022 by IntechOpen
IntechOpen is the global imprint of INTECHOPEN LIMITED, registered in England and Wales, registration number: 11086078, 5 Princes Gate Court, London, SW7 2QJ, United Kingdom
Printed in Croatia

British Library Cataloguing-in-Publication Data
A catalogue record for this book is available from the British Library

Additional hard and PDF copies can be obtained from orders@intechopen.com

Quantum Computing and Communications
Edited by Yongli Zhao
p. cm.
Print ISBN 978-1-83968-133-2
Online ISBN 978-1-83968-134-9
eBook (PDF) ISBN 978-1-83968-135-6

Meet the editor

Yongli Zhao received a Ph.D. from Beijing University of Posts and Telecommunications (BUPT), Beijing, China, in 2010. He is currently a full professor at the School of Electronic Engineering, BUPT. In 2016–2017, Dr. Zhao was a visiting associate professor at the University of California, Davis. He is a member of the Institution of Engineering and Technology (IET), Institute of Electrical and Electronics Engineers (IEEE), and OSA. He has published more than 400 international journal and conference papers. His current research focuses on optical network security and quantum communications.

Contents

Preface

The theory of quantum computing and communication has made remarkable progress over the years and has become a significant scientific discipline. It encompasses many concepts related to quantum information technologies, including quantum algorithms, quantum computers, quantum code, post-quantum cryptography, quantum key distribution, and quantum teleportation.

This book surveys the field of quantum computation and quantum communication from a fresh perspective, discussing its representative technologies and the latest research. The introductory chapter provides an overview of the topic and the book. Chapter 2, "Quantum Algorithms for Nonlinear Equations in Fluid Mechanics", gives a narrative tutorial on quantum algorithms. Chapter 3, "A Novel Three-Input XOR Gate Based on Quantum Dot-Cellular Automata with Power Dissipation Analysis", presents the "quantum code" notation and restates many of the examples and results of the preceding chapter. Chapter 4, "Topology in Photonic Discrete-Time Quantum Walks: A Comprehensive Review", provides a comprehensive review of photonic implementations of discrete-time quantum walks in the spatial and temporal domains. Chapter 5, "Introduction to Quantum Computing", explains how a quantum computer might be built. Chapter 6, "Multipoint-Interconnected Quantum Communication Networks", discusses some quantum applications, including quantum key distribution and teleportation along with related development and research.

We would like to thank Dr. Rene Steijl, Professor Ismail Gassoumi, Professor Graciana Puentes, and Professor Surya Teja Marella for their help in writing some chapters. We also received help, support, and encouragement from many other individuals and institutes. We would also like to thank those physicists and computer scientists who have developed the field of quantum computation and quantum communication.

Yongli Zhao
Beijing University of Posts and Telecommunications,
Beijing, China

Chapter 1

Introductory Chapter: Quantum Computing and Communications

Yongli Zhao, Yazi Wang and Xiaosong Yu

1. Introduction

1.1 The origin of quantum information

Quantum mechanics' establishment and development triggered the first wave of quantum technology in the twentieth century. With the regulation and observation of microphysical quantity as the main feature of understanding and grasping the microphysical phenomena and laws, quantum information based on the principles of quantum mechanics was born. Quantum information, a new information method, that calculates, encodes, and transmits the physical information contained in the "state" of a quantum system. The most common unit of quantum information is the qubit, that is, intrinsically linked to each other and can be any combination of 0 and 1 simultaneously.

2. The development history of quantum information

Quantum information technologies aim to use the natural characteristic of the atomic scale to accomplish tasks that cannot be achieved with existing technologies and use the characteristic of measuring or observing a quantum system to change the quantum information fundamentally. These technologies rely on qubits. Meanwhile, scientists are creating physical qubits from a variety of particles, such as atoms or light particles, or objects that mimic them, such as superconducting circuits. Scientists manipulate the quantum properties of each qubit and entangle multiple qubits with each other to create quantum technology from these qubits. These functions support two potential transformative applications, that is, quantum computing and quantum communications. However, quantum information is fragile and can be irreversibly lost through interactions with the environment, a process known as decoherence. Quantum error correction techniques have been proposed and proven, but are challenging to implement. Based on these, researchers began to explore the application of quantum information to quantum technologies in the twentieth century.

- In 1959, researcher Richard Feynman believed that manipulating matter at the atomic scale is possible, meaning that certain types of computation can be done more efficiently on quantum systems than on classical [1].

- In 1972, researchers showed that one qubit measurement could affect other qubits, which is the first proof of entanglement [2].

- In 1981, researchers observed that it might not be possible to effectively simulate the evolution of quantum systems on classical computers and proposed a basic model of quantum computing.

- In 1984, researchers described a quantum key distribution scheme; in this scheme, eavesdroppers had a high probability of being detected when they tried to monitor an encrypted key exchange that used qubits to transmit information. The scheme, often referred to as BB84, is considered the first quantum cryptography protocol [3].

- In 1987, researchers found the property necessary for photon entanglement by measuring the time interval between two photons and found these two photons were indistinguishable from each other [4].

- In 1991, researchers extended the BB84 protocol and introduced a different method of quantum key distribution that contains entanglement [5].

- In 1994, the well-known American physicist Peter Shor proposed the well-known quantum algorithm, which is the Shor quantum decomposition algorithm. The Shor quantum decomposition algorithm is based on the Deutsch-Jozsa algorithm [6], following the laws and theories of quantum mechanics [7].

- In 1998, researchers demonstrated through principle experiments that quantum error correction is possible, which is necessary for cost-effective quantum computing and communication because excessive noise can destroy quantum information.

In the twenty-first century, the theory and development of quantum computing and communications puts this significance on a firm footing and leads to some new profound and exciting insights into the natural world. From 2000 to 2005, a variety of time-efficient quantum algorithms were proposed, such as the semi-product groups [8–10], the near-Hamiltonian groups [11], the normal subgroups [12, 13], the almost Abelian groups [14]. In 2006, Hayashi et al. [15] proposed the first quantum network coding scheme, which realized the cross transmission of two qubits in a full quantum channel butterfly network. Due to the constraints of quantum properties, such as the quantum unclonable theorem, this scheme cannot achieve lossless quantum transmission, that is, the fidelity is less than 1. In 2012, Satoh et al. [16] designed a novel quantum network coding scheme using quantum repeaters. In 2014, Nishimura [17] summarized the current state of quantum network coding, discussing the nature of quantum network coding schemes using entangled resources to communicate with classical. In 2020, Wu et al. [18] proposed a continuous-variable quantum network coding scheme based on a butterfly-shaped network model.

3. Quantum revolution with quantum computing and communication

Among these, some quantum computing and communication technologies are available for use; for example, quantum computer, quantum cryptography, teleportation, and quantum error correction. Quantum computer is the physical platform that realizes quantum computing to encode qubits so that different qubits can be coupled in a controllable manner and have a certain resistance to the influence of the noise environment. The main quantum computer solutions currently developed include superconductivity, ion traps, quantum dots, topologies, and diamond color centers. Quantum cryptography can take advantage of quantum states to enable classical information to be transmitted securely. Teleportation achieves reliable transmission of quantum states by using entanglement. Quantum error correction keeps the possibility of maintaining quantum coherence when an irreversible noise process exists.

At present, the world is undergoing a new round of "quantum revolution". Quantum computing and communication technologies are accelerating breakthroughs in key technologies. In the new stage, some technologies are gradually being integrated with the system. Breakthroughs have been made in key technologies, such as integration, engineering, and networking. The integration of quantum communication with classical communication networks, multi-bit operation, and computing of quantum computers show the application prospects of quantum information in the industry.

4. Brief introduction of the book

In this book, we will introduce some fundamental quantum computing and communication technologies that will form the basis for much of what follows. After this brief introduction, we will review the basic conception and relevance of quantum algorithms, quantum computer, quantum code, post-quantum cryptography, quantum key distribution, and quantum teleportation respectively in detail.

Author details

Yongli Zhao*, Yazi Wang and Xiaosong Yu
Beijing University of Posts and Telecommunications, Beijing, China

*Address all correspondence to: yonglizhao@bupt.edu.cn

References

[1] Feynman R. There is plenty of room at the bottom: An invitation to enter a new field of physics. In: Lecture at American Physical Society Meeting. Vol. 29. 1959

[2] Freedman SJ, Clauser JF. Experimental test of local hidden-variable theories. Physical Review Letters. 1972;**28**(14):938-941

[3] Bennett CH, Brassard G. Quantum cryptography: Public key distribution and coin tossing. In: International Conference on Computers, Systems, and Signal Processing. 1984. pp. 175-179

[4] Hong CK, Ou ZY, Mandel L. Measurement of subpicosecond time intervals between two photons by interference. Physical Review Letters. 1987;**59**(18):2044-2046

[5] Eckert AK. Quantum cryptography based on Bell's theorem. Physical Review Letters. 1991;**67**(6):661-663

[6] Deutsch D. Quantum theory, the church–Turing principle and the universal quantum computer. In: Proceedings of the Royal Society of London, A, Mathematical and Physical Sciences. 1985

[7] Shor P. Algorithms for quantum computation: Discrete logarithms and factoring. In: 35th Annual Symposium on Foundations of Computer Science. 1994. pp. 124-134

[8] Kuperberg G. A subexponential-time quantum algorithm for the dihedral hidden subgroup problem. SIAM Journal on Computing. 2005;**35**(1):170-188

[9] Moore C, Rockmore D, Russell A, et al. The power of basis selection in Fourier sampling: The hidden subgroup problem in affine groups. In: Proc. SODA. 2004

[10] Yoshifumi I, Le Gall F. Efficient quantum algorithms for the hidden subgroup problem over a class of semi-direct product groups. arXiv preprint quant-ph/0412033. 2004

[11] Gavinsky D. Quantum solution to the hidden subgroup problem for poly-near-Hamiltoniangroups. Quantum Information and Computation. 2004;**4**:229-235

[12] Hallgren S, Russell A, Ta-Shma A. Normal subgroup reconstruction and quantum computation using group representations. SIAM Journal on Computing. 2003;**32**(4):916-934

[13] Ivanyos G, Magniez F, Santha M. Efficient quantum algorithms for some instances of the non-abelian hidden subgroup problem. In: SPAA. New York: ACM Press; 2001. pp. 263-270

[14] Grigni M, Schulman L, Vazirani M, et al. Quantum Mechanical Algorithms for the Nonabelian Hidden Subgroup Problem. In: Proceedings of the thirty-third annual ACM symposium on Theory of computing; 2001. pp. 68-74

[15] Hayashi M et al. Quantum network coding. In: Annual Symposium on Theoretical Aspects of Computer Science. Berlin, Heidelberg: Springer; 2007

[16] Satoh T, Le Gall F, Imai H. Quantum network coding for quantum repeaters. Physical Review A. 2012;**86**(3):032331

[17] Nishimura H. Quantum network coding and the current status of its studies. In: 2014 International Symposium on Information Theory and its Applications. IEEE; 2014

[18] Qu Z et al. Continuous-variable quantum network coding protocol based on butterfly network model. International Journal of Sensor Networks. 2020;**32**(2):69-76

Chapter 2

Quantum Algorithms for Nonlinear Equations in Fluid Mechanics

Rene Steijl

Abstract

In recent years, significant progress has been made in the development of quantum algorithms for linear ordinary differential equations as well as linear partial differential equations. There has not been similar progress in the development of quantum algorithms for nonlinear differential equations. In the present work, the focus is on nonlinear partial differential equations arising as governing equations in fluid mechanics. First, the key challenges related to nonlinear equations in the context of quantum computing are discussed. Then, as the main contribution of this work, quantum circuits are presented that represent the nonlinear convection terms in the Navier–Stokes equations. The quantum algorithms introduced use encoding in the computational basis, and employ arithmetic based on the Quantum Fourier Transform. Furthermore, a floating-point type data representation is used instead of the fixed-point representation typically employed in quantum algorithms. A complexity analysis shows that even with the limited number of qubits available on current and near-term quantum computers (<100), nonlinear product terms can be computed with good accuracy. The importance of including sub-normal numbers in the floating-point quantum arithmetic is demonstrated for a representative example problem. Further development steps required to embed the introduced algorithms into larger-scale algorithms are discussed.

Keywords: partial differential equations, fluid mechanics, nonlinear equations, quantum Fourier transform, floating-point arithmetic

1. Introduction

Quantum computing [1] and quantum communication are research areas that have seen significant developments and progress in recent years, as is apparent from the work covered in this book. In this chapter, the focus is on the development of quantum algorithms for solving nonlinear differential equations, highlighting key challenges that arise from the non-linearity of the equations to be solved. For this application of quantum computing, progress has so far been relatively limited and in this work, a promising approach to deriving efficient quantum algorithms is proposed. Although the focus is on non-linear equations related to fluid mechanics, the approach put forward here is applicable to a much wider range of problems.

IntechOpen

Furthermore, in developing the proposed method, efficient quantum circuits involving floating-point arithmetic were created, in contrast to the more commonly used fixed-point arithmetic employed in a range of quantum algorithms. This aspect of the work described here should also be useful for a wider audience. In this work, the development of quantum algorithms for the nonlinear governing equations for fluid mechanics is described with a particular focus on representing the non-linear product terms in the equations. A key aspect of the derived quantum circuits in the present work is the (temporary) representation of the solution in the computational basis, along with the the use of a floating-point data representation in the arithmetic operations. The quantum circuits for obtaining the non-linear product terms are new developments and form the main contribution of this work. In recent years, a small number of works have considered quantum computing applications to fluid mechanics [2–8]. A brief review of this previous work will be presented in Section 2 and will provide context to the proposed approach. Related work on algorithms with representation in the computational basis is reviewed in this chapter. This chapter is structured as follows. Section 2 describes the background to the current work. Section 3 reviews the key challenges related to treating nonlinear differential equations in a quantum computing context, followed by a discussion of the nonlinear governing equations in fluids dynamics in Section 4. Section 5 then describes how nonlinear terms in governing equations can be evaluated in quantum algorithms using the computational basis. Section 6 and Section 7 discuss the quantum circuits used for computing the square of a floating-point number and the multiplication of two floating-point numbers, respectively. The simulation and verification of the derived quantum circuits is presented in Section 8. The complexity of the circuits is analyzed in Section 9. Finally, conclusions from this work and suggestions for further work are presented in Section 10.

2. Background of present work

For a small number of applications, quantum algorithms have been developed that display a significant speed-up relative to classical methods. Computational quantum chemistry is proving to be one of the key areas of application. Important developments for a wider range of applications include quantum algorithms for linear systems [9, 10] and the Poisson equation [11]. Applications to computational science and engineering problems beyond quantum chemistry have only recently begun to appear [4–6, 12–14]. Despite this research effort, progress in defining suitable engineering applications for quantum computers has been limited.

Significant progress has been made in recent years in the development of quantum algorithms for linear ordinary differential equations (ODEs) as well as linear partial differential equations (PDEs) [15–19]. However, in contract to this progress for linear equations, there has not been similar progress in the development of quantum algorithms for nonlinear ODEs and nonlinear PDEs. An early work by Leyton and Osborne [20] presented an innovative and highly ambitious algorithm. However, the computational complexity of this work involves exponential dependency on the time interval used in the time integration. A small number of more recent works have addressed nonlinear differential equations and typically algorithms for very specific problems were obtained [8]. Therefore, much research work is needed into quantum algorithms for a wider range of nonlinear problems.

Early work in quantum computing relevant to the field of Computational Fluid Dynamics (CFD) mainly involved the work on quantum lattice-gas models by

Yepez and co-workers [2, 3]. This work typically used type-II quantum computers, consisting of a large lattice of small quantum computers interconnected in nearest neighbor fashion by classical communication channels. In contrast to these quantum lattice-gas based approaches, the present study focuses on quantum algorithms designed for near-future 'universal' quantum computers. The potential of quantum computing in the context of direct numerical simulation of flows was reviewed recently by Griffin et al. [7], showing that a number of further developments are needed to make this approach viable.

Typically, there are two methods of encoding the result of a quantum algorithm: encoding within the computational basis of the quantum state and encoding within the amplitudes of the quantum state. The widely-used Quantum Fourier Transform (QFT) uses the second approach. The QFT with complexity $O(log^2(N))$ for problem size N has exponential speed-up compared to the classical fast Fourier transform (complexity $O(NlogN)$) and plays an important role in quantum computation as an essential part of many quantum algorithms. The exponential speed-up realized is due to superposition and quantum parallelism. However, in same quantum algorithms, the Fourier coefficients may be needed in the computational basis [21].

Here, the two different encoding methods are illustrated using the discrete Fourier Transform (DFT). The QFT performs the DFT in terms of amplitudes as,

$$\sum_{j=0}^{N-1} x_j|j\rangle \rightarrow \sum_{k=0}^{N-1} y_k|k\rangle \tag{1}$$

The QFT performs a DFT on a list of complex numbers, and the result is stored as amplitudes of a quantum state vector. In order to extract the individual Fourier components, measurements need to be performed on the quantum state vector. Therefore, the QFT is not directly useful for determining the Fourier-transformed coefficients of the input state. However, the QFT is widely used as a subroutine in larger algorithms. In contrast to the amplitude encoding in Eq. (1), Zhou et al. [21] presented a quantum algorithm computing the Fourier transform in the computational basis (termed QFTC). This quantum algorithm encodes Fourier coefficients with fidelity $1-\delta$ and digit accuracy ε for each Fourier coefficient. Its time complexity depends polynomially on $log\,(N)$, and linearly on $1/\delta$ and $1/\varepsilon$. The QFTC, enables the Fourier-transformed coefficient to be encoded in the computational basis as follows,

$$|k\rangle|0\rangle \rightarrow |k\rangle|y_k\rangle \tag{2}$$

where y_k corresponds to the fixed-point binary representation of $y_k \in (-1, 1)$ using two's complement format. In the algorithm proposed by Zhou et al. [21], the input vector $\vec{x}$ is provided by an oracle O_x such that,

$$O_x|0\rangle = \sum_{j=0}^{N-1} x_j|j\rangle \tag{3}$$

which can be efficiently implemented if $\vec{x}$ is efficiently computable or by using the qRAM that takes complexity $log\,(N)$ under certain conditions [21]. Comparing Eq. (1) and Eq. (2), it is clear that encoding in the computational basis requires a number of additional qubits depending on the required fixed-point representation.

3. Nonlinear problems on quantum computers

An early work by Leyton and Osborne [20] introduced a quantum algorithm to solve nonlinear differential equations with an unfavorable complexity. Since then, very few works have considered quantum algorithms for nonlinear equations. In contrast, algorithms for linear differential equations have continued to receive significant attention. As an example, advanced quantum spectral methods for differential equations were published recently by Childs and Liu [19].

A key contributing factor to the limited progress in algorithms for non-linear problems is the inherent linearity of quantum mechanics. For quantum algorithms encoding information as amplitudes of a quantum state vector, nonlinear (product) terms cannot be obtained by multiplying these amplitudes by themselves, as a result of the no-cloning theorem that prohibits the copying of an arbitrary quantum state. Furthermore, all quantum-gate operations (with the exception of measurements) in the quantum-circuit model used here need to be unitary and reversible. These requirements add further challenges to representing nonlinear terms when using the amplitude-based encoding approach. Specifically, in a normalized quantum state vector all amplitudes in the vector are ≤ 1 (unless only a single amplitude is non-zero), therefore an operator performing products of the amplitudes cannot be unitary since the resulting quantum state vector will no longer have a unit norm.

One possible way around these problems associated with nonlinear terms would be a hybrid quantum-classical approach where the nonlinear products are computed on a classical computer. However, due to the complexity introduced by measuring the quantum state (needed before each transfer of information to the classical computer) and the cost of (re-)initialization of the quantum computer with the result of these products, this is not a promising line of development. It is highly unlikely to lead to a quantum speed-up. Recently, Variational Quantum Computing (VQC) was introduced as an effective hybrid classical-quantum approach [22, 23], firstly for applications in quantum chemistry and more recently for a wider range of linear and nonlinear problems [24]. The VQC approach constructs the required solution from a layered network, as illustrated in **Figure 1**. As shown in **Figure 1(a)**, multiple layers are used (4 in the illustration), each taking as input multiple qubits (6 in example shown). Using depth 5 in the example, the quantum circuits defined

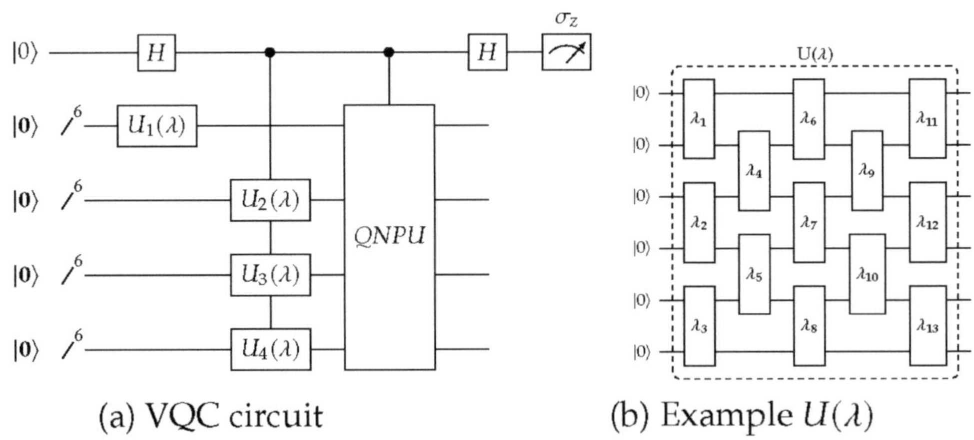

Figure 1.
Illustration of the Variational Quantum computing (VQC) approach (adapted from Lubasch et al. [24]).

by $U(\lambda)$ involve 13 two-qubit gates as shown in **Figure 1(b)**. Each of these gates has a parameter $\lambda_i \in [1, 13]$ associated with it. A classical computer is used to create optimized parameters λ employing an iterative approach that takes the measured state of the ancilla qubit as input. A further key part of the approach is the problem-specific Quantum Nonlinear Processing Unit (QNPU). Recently, Lubasch et al. [24] published an example for the QNPU for the nonlinear Burgers equation. In applications of the VQC approach, the efficiency strongly depends on the choice of the number of parameters λ used in $U(\lambda)$. The work by Lubasch et al. [24] showed that exponential speed-up is only possible if the depth of $U(\lambda)$ scales with the number of qubits and not with the overall problem size. It is clear that the proposed VQC approach is an important development toward QC applications to nonlinear problems. It therefore constitutes a leading candidate for applications to fluid dynamics. However, it is also clear that further investigation is needed to further assess its suitability for a range of applications.

4. Nonlinear governing equations in fluid mechanics

The Navier–Stokes equations for an incompressible, Newtonian fluid can be written as,

$$\frac{\partial \mathbf{U}}{\partial t} + \mathbf{U} \cdot \nabla_{\mathbf{x}} \mathbf{U} = -\frac{1}{\rho} \nabla_{\mathbf{x}} p + \nu \Delta \mathbf{U} \quad ; \quad \nabla_{\mathbf{x}} \cdot \mathbf{U} = 0 \tag{4}$$

where $\mathbf{U}$, p, ρ and ν are the velocity, pressure, density and kinematic viscosity, respectively. $\mathbf{x}$ denotes the coordinate in space. The second term on the right-hand side of Eq. (4) is the nonlinear convection term that poses a key challenge to developing efficient quantum algorithms for the Navier–Stokes equations. Efficient quantum algorithms for linear convection equations discretized on regular Cartesian meshes with periodic boundary conditions have been devised in recent years [6]. When studying numerical methods for the Navier–Stokes equations, it is often useful to switch to Burgers' model equation, to obtain a single nonlinear partial differential equation that retains a nonlinear convection term similar to the Navier–Stokes equations. Using the VQC approach, Lubasch and co-workers recently published example quantum circuits to model the Burgers equation [24]. Griffin et al. [7] discuss two approaches for treating the nonlinear term in the Navier–Stokes equations: the VCQ approach of Lubasch et al. [24] and a linearized approach. These authors conclude that, at present, VQC represents the most promising approach for Navier–Stokes equations. Their study also highlights that much further research work is needed to create efficient algorithms for fluid dynamics applications. It is relatively easy to show that the linearization approach to solving non-linear governing equations on a Quantum Computer is generally unfeasible. In applying linearization to nonlinear governing equations, the idea is to use a linearization about the present state of the solution, and then advance this linearized problem in time. This creates a linearization error, which is small if the time step is small. However, even if this linearization error can be tolerated, the linearization approach is problematic in a quantum computing context. This is due to the need for repeated measuring of the quantum state (so that the gates that implement the linear operator may be updated with the current solution) and repeated re-initialization of the quantum state. The complexity associated with repeated measuring and re-initialization is so large that any benefit of a quantum algorithm over a classical algorithm is very likely to vanish.

The development of quantum algorithms for fluid dynamics is clearly at a very early stage and therefore it is essential that different approaches are considered.

5. Representing nonlinear terms in computational basis

In the present work, an alternative approach to introducing the nonlinear terms of nonlinear differential equations into a quantum algorithm is investigated. Specifically, the assumption is made that in a large-scale quantum algorithm for the solution of the nonlinear (partial) differential equations, the solution is encoded in terms of amplitude in the quantum state vector, i.e. the approach used in a wide range of algorithms including the QFT. Then, for the nonlinear terms of the equations, the following steps are suggested. First, within the larger quantum algorithm, a quantum algorithm is embedded that converts the solution from the quantum-amplitude representation to a representation in the computational basis. Recently, quantum algorithms for this 'analog-to-digital conversion' were published by Mitarai et al. [25]. Using the representation of the solution in the computational basis, the required nonlinear terms are then efficiently evaluated using quantum circuits presented later in this chapter. Once computed, a conversion back to quantum-amplitude representation is to be used, enabling the rest of the quantum algorithm to proceed. For this 'digital-to-analog' conversion, quantum algorithms were recently studied and published by SaiToh [26]. For the representation in the computational basis, a fixed-point approach is typically employed to represent real or complex numbers in quantum algorithms. The number of additional qubits required when using computational-basis encoding depends directly on the number of qubits required to represent the real and complex numbers needed in the algorithm. In the present work, a different approach is put forward: instead of using fixed-point arithmetic, a floating-point representation is used.

In the literature, quantum arithmetic using floating-point numbers has received very little attention so far. Haener et al. [27] described an investigation into quantum circuits for floating-point addition and multiplications and compared automatically generated circuits from Verilog implementations with hand-crafted optimized circuits. Their study provides evidence that floating-point arithmetic is a viable candidate for use in quantum computing, at least for typical scientific applications, where addition operations usually do not dominate the computation. Following on from these conclusions, the present work investigates the use of floating-point arithmetic as part of evaluating nonlinear terms in the computational basis.

5.1 Previous works on algorithms in computational basis

Quantum arithmetic in the computational basis constitutes an important component of many quantum algorithms, and as a result reversible implementations of algebraic functions (addition, multiplication, inverse, square root, etc.) have been widely studied. In contrast, there is relatively little work on quantum algorithm implementation of higher-level transcendental functions, such as logarithmic, exponential, trigonometric and inverse trigonometric functions. Examples of applications of trigonometric and inverse trigonometric functions in the computational basis can be found in the famous HHL algorithm [9] and in the state preparation algorithm introduced by Grover and Rudolph [28]. More recently, a quantum algorithm for approximating the QR decomposition of a $N \times N$ matrix in the computational basis was published by Ma et al. [29], with polynomial speed-up over the best classical algorithm.

5.2 Fixed-point and floating-point arithmetic

A fixed-point number held in an n_q qubit register can be defined as the following quantum state,

$$|w\rangle = \overbrace{\left|w^{(n_{int}-1)}\right\rangle \otimes \left|w^{(n_{int}-2)}\right\rangle \otimes \ldots \otimes \left|w^{(0)}\right\rangle}^{integer} \otimes \overbrace{\left|w^{(-1)}\right\rangle \otimes \ldots \otimes \left|w^{(n_{int}-n_q)}\right\rangle}^{fractional} \quad (5)$$

where $w^{(j)} \in 0, 1, j = n_{int} - n_q, n_{int} - n_q + 1, \ldots, 0, \ldots, n_{int-1}$ [30]. This state represents the number $w = \sum_j w^{(j)} 2^j$. The n_{int} leftmost qubits are used to represent the integer part of the number and the remaining $n_{frac} = n_q - n_{int}$ qubits represent its fractional part. In this example, no sign qubit is used so that only positive numbers can be represented (for most applications an additional sign qubit would be required). Since fewer than n_q bits may suffice for the representation of the input, a number of the leftmost qubits in the register may be set to $|0\rangle$. Clearly, the fixed point system is very limited in terms of the size of the numbers it can store. Therefore, soon after computers were introduced for numerical computing the switch to floating-point arithmetic was made. In a computer implementation of a floating point number with base 2, a non-zero signed number x, defined through a normalized representation, is expressed as,

$$x = \pm S \times 2^E, \quad \text{where} \quad 1 \leq S < 2 \quad (6)$$

where the numbers S and E are the mantissa and the exponent, respectively. The binary expansion of the mantissa is

$$S = (b_0.b_1b_2b_3\ldots)_2 \text{ with } \quad b_0 = 1 \quad (7)$$

Here, it is important to note that always $b_0 = 1$ for non-zero numbers in a normalized representation. This will be used in the present work to achieve savings in the number of required qubits, as detailed later. In the binary representation, the bits following the binary point are the fractional part of the mantissa. Once floating-point numerical computation on classical computers became commonplace, the industry standard IEEE 754 was introduced [31]. A similar standard for floating-point representations on a quantum computer does not yet exist, but is desirable [30]. A key feature of the IEEE standard is that it requires correctly rounded operations: correctly rounded arithmetic operations, correctly rounded remainder and square root operations and correctly rounded format conversions. Typically, rounding to the nearest floating pointing number available in the destination (output register) is used. In the quantum circuits in the present work, rounding down to nearest is used, for reasons of simplicity. Detailed analysis of quantum-circuits developed here for squaring and multiplication operations shows that 'correctly' rounding to nearest involves a significant increase in circuit complexity (i.e. using quantum equivalents of guard and sticky bits, that are well established in arithmetic on classical computers [31]). A key aspect of the IEEE 754 that has been incorporated in the present work is the definition of *sub-normal numbers*. To illustrate the concept of subnormal numbers, the IEEE 754 standard representation of single format numbers using a 32-bit word is considered. The first bit is the sign bit, followed by 8 bits representing the exponent. Then, 23 bits are used to store a 24-bit representation of the mantissa, i.e. b_0 is not stored. Numbers with exponent bits $(00000000)_2 = (0)_{10}$ and $(11111111)_2 = (255)_{10}$ are defined special cases. The smallest normalized number is $(1.0 \ldots 0)_2 \times 2^{-126} = 2^{-126}$. Sub-normal numbers are used to represent smaller numbers, i.e. in this case the exponent field has a zero bit string but the fraction field has a nonzero bit string. Zero is represented with a zero bit string for the fractional field. For all subnormal numbers, the (00000000) used for the exponent represents 2^{-126} and by using the 23 fractional field bits, equally-spaced numbers in the range $(0.00 \ldots 01)_2 \times 2^{-126}$ (with 22 zero bits after the binary point) to $(0.11 \ldots 11)_2 \times 2^{-126}$ (with 23 one bits after the binary point) are encoded.

5.3 Quantum floating-point format used in present work

Based on the floating point representation defined in the IEEE standard, the present work introduces a floating-point system with fewer bits (i.e. qubits in this case) than the 32 used for single format numbers. This is the direct result of the limited number of qubits available on current and near-term quantum computers. To optimize the range of floating-point numbers that can be represented with the approach used here, the following key aspects of the IEEE standard were adopted:

- For the mantissa only the fractional part is stored,
- Exponent bit strings $(00 \ldots 00)_2$ and $(11 \ldots 11)_2$ are used for special cases, i.e. dealing with 0, subnormal numbers as well as cases with overflow,
- The remaining range of exponent bit strings is used for a range of exponential centred around $2^0 = (01 \ldots 11)_2$,
- Sub-normal numbers are used to extend the range of small numbers,
- Rounding down to nearest is used as rounding mode,
- Only unsigned numbers are considered for simplicity. Signed numbers can easily be obtained by adding a further 'sign' qubit.

In this work, a floating-point number is represented as an $n_q = N_M + N_E$ quantum register. In the quantum-circuit implementation, the most significant (left-most) mantissa qubit is not stored, using the hidden-bit approach used in the IEEE 754 standard. Therefore, $N_M - 1$ qubits define the fractional part of the mantissa in the developed quantum circuits. N_E defines the number of qubits used to define the exponent. In the following, examples with $N_E = 3$ and $N_E = 4$ and $N_M \in [3, 5]$ are considered. For $N_E = 3$, the number 1.00 is defined by $|00|011\rangle$ when $N_M = 3$. Similarly, $|000|0111\rangle$ defines the number 1.000 for $N_M = 4$ and $N_E = 4$. For $N_E = 3$, the smallest normalized number that can be represented is $1/4$ independent of the

$N_M = 3$	$N_M = 4$	$N_M = 5$
$\|011\|000\rangle = 3/16$	$\|0111\|000\rangle = 7/32$	$\|01111\|000\rangle = 15/64$
$\|010\|000\rangle = 1/8$	$\|0110\|000\rangle = 3/16$	$\|01110\|000\rangle = 7/32$
$\|001\|000\rangle = 1/16$	$\|0101\|000\rangle = 5/32$	$\|01101\|000\rangle = 13/64$
	$\|0100\|000\rangle = 1/8$	$\|01100\|000\rangle = 3/16$
	$\|0011\|000\rangle = 3/32$	$\|01011\|000\rangle = 11/64$
	$\|0010\|000\rangle = 1/16$	$\|01010\|000\rangle = 5/32$
	$\|0001\|000\rangle = 1/32$	$\|01001\|000\rangle = 9/64$
		$\vdots$
		$\|00010\|000\rangle = 1/32$
		$\|00001\|000\rangle = 1/64$

Table 1.
Sub-normal numbers for floating-point numbers with 3 qubits as exponential.

$N_M = 4$	$N_M = 5$
$\|0111\|0000\rangle = 7/512$	$\|01111\|0000\rangle = 15/1024$
$\|0110\|0000\rangle = 3/256$	$\|01110\|0000\rangle = 7/512$
$\|0101\|0000\rangle = 5/512$	$\|01101\|0000\rangle = 13/1024$
$\|0100\|0000\rangle = 1/128$	$\|01100\|0000\rangle = 3/256$
$\|0011\|0000\rangle = 3/512$	$\|01011\|0000\rangle = 11/1024$
$\|0010\|0000\rangle = 1/256$	$\|01010\|0000\rangle = 5/512$
$\|0001\|0000\rangle = 1/512$	$\|01001\|0000\rangle = 9/1024$
	$\vdots$
	$\|00010\|0000\rangle = 1/512$
	$\|00001\|0000\rangle = 1/1024$

Table 2.
Sub-normal numbers for floating-point numbers with 4 qubits as exponential.

number mantissa qubits. Then, exponent state $|000\rangle$ defines zero and sub-normal numbers, as shown in **Table 1** for $N_M = 3$, $N_M = 4$ and $N_M = 5$.

Similarly, using 4 qubits for the exponent ($N_E = 4$) means that the smallest normalized number is $1/64$. For $N_M = 4$ and $N_M = 5$, **Table 2** shows the corresponding sub-normal numbers.

In line with the IEEE 754 standard, exponent state $|1 \dots 1\rangle$ denotes numbers for which an overflow has occurred. For $N_E = 3$, the largest normalized number available is $|11 \dots 1|110\rangle$ which equates to 14 and 15 for $N_M = 3$ and $N_M = 4$, respectively. Similarly, for $N_E = 4$, the largest normalized number available is $|11 \dots 1|1110\rangle$ which equates to 240 and 248 for $N_M = 4$ and $N_M = 5$, respectively.

6. Quantum circuits for squaring floating-point numbers

For a floating-point number defined by N_M mantissa and N_E exponent bits, a total of $N_M - 1 + N_E$ qubits is needed to define the state in the quantum circuits introduced here. An example with $N_M = 3$ and $N_E = 3$ will now considered, using registers $|imb1|imb0\rangle$ and $|ieb2|ieb1|ieb1\rangle$ to define the fractional part of the mantissa and the exponent of the input number, respectively. For the multiplication operation described later a second input floating-point number is defined using $|ima1|ima0\rangle$ and $|iea2|iea1|iea1\rangle$. The output of the squaring and multiplication operations is a floating-point number r defined by $|imr1|imr0\rangle$ and $|ier2|ier1|ier1\rangle$ (initialized at $|\mathbf{0}\rangle$). In addition to the input and output registers, the quantum circuits will need additional qubits to hold results of intermediate results, e.g. for $N_M = 3$ a 6-qubit sub-register $|imp5| \ldots |imp0\rangle$ is used. To facilitate the quantum-multiplication operations, a further ancilla qubit $|a0\rangle$ is used. For quantum circuits without measures to deal with sub-normal numbers and overflow, the quantum state for $N_M = 3$ and $N_E = 3$ is defined in a $2 \times (N_M - 1 + N_E) + 2 \times N_M + 1 = 17$-qubit register

$$|ieb2|ieb1|ieb0|imb1|imb0|a0|imp5| \ldots |imp0|ier2|ier1|ier0|imr1|imr0\rangle \qquad (8)$$

For $N_M = 4$ and $N_E = 4$, the required number of qubits increases to $2 \times (N_M - 1 + N_E) + 2 \times N_M + 1 = 23$. The quantum circuit performing the squaring operation for $N_M = 3$ and $N_E = 3$ is detailed here as example (in realistic applications $N_M > 3$ will typically be needed). **Figure 2** shows the quantum circuit used in the first step of computing the square of a quantum floating point with $N_M = 3$ and $N_E = 3$. This step involves computing the square of the mantissa, with this result temporarily stored in $|imp5| \ldots |imp0\rangle$. In this circuit, $QFT6$ prepares this temporary register for the three subsequent product steps denoted by P_1, P_2 and $P3$, involving doubly-controlled phase operations. Specifically, three-qubit gates are used applying a phase rotation conditional on state of $|a0\rangle$ and either $|imb1\rangle$ or $|imb0\rangle$. The P_i steps are controlled-summation operations in the shift-and-add approach to computing the products, i.e. the circuits in P_i are derived from quantum adders

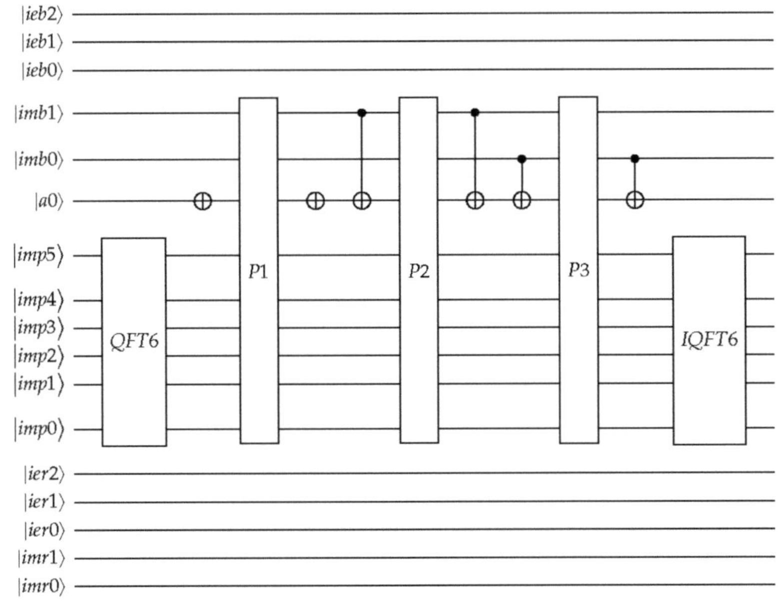

Figure 2.
Quantum circuit used to compute square of mantissa (for $N_M = 3$ and $N_E = 3$).

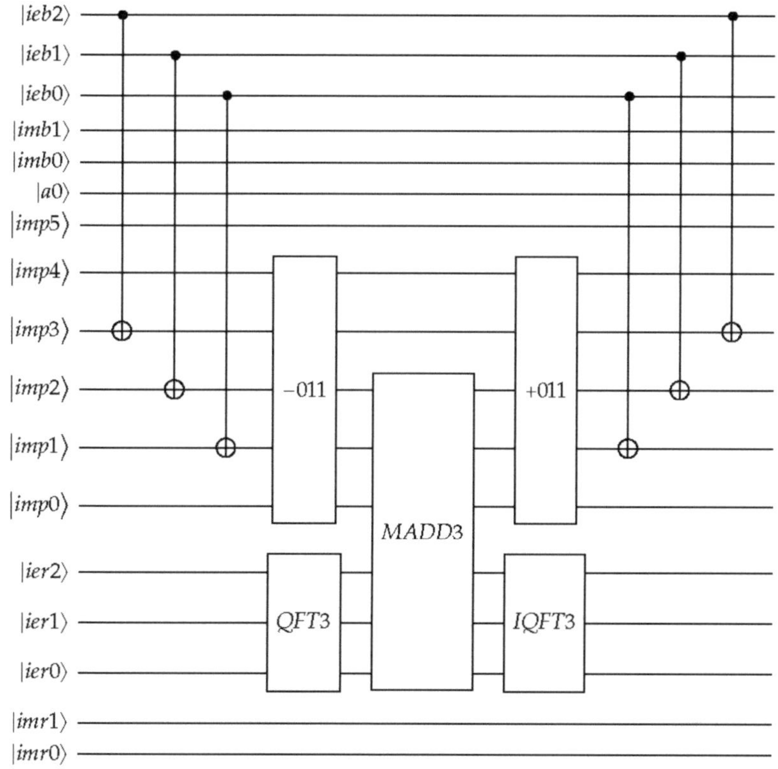

Figure 3.
Quantum circuit used to obtain exponent for squaring operation ($N_M = 3$ and $N_E = 3$).

controlled by an additional qubit. Once the controlled phase changes in the circuits P_1, P_2 and P_3 have been applied, inverse *QFT*6 on $|imp5| \dots |imp0\rangle$ creates the desired output state. In case the square of the mantissa ≥ 2, i.e. $|imp5 = 1\rangle$, the result exponent needs to be incremented by 1. This is achieved by apply a controlled-NOT to $|ier0\rangle$ (which was initialized at $|0\rangle$) with $|imp5\rangle$ as control. In the next step, result mantissa qubits $|imr1|imr0\rangle$ are set using temporary results in $|imp5| \dots |imp0\rangle$, where the required gate operations are conditional on the state of $|imp5\rangle$. Then the steps shown in **Figure 2** are 'uncomputed' so that the sub-register $|imp5| \dots |imp0\rangle$ is set to $|\mathbf{0}\rangle$ again. The next step is illustrated in **Figure 3**, where the output exponent is obtained. This step involves the initialization of the temporary register $imp3| \dots |imp0$ with $2 \times E_b$ (i.e. twice the input exponent). Then, the bias of $(011)_2 = 3$ is removed (denoted by -011). This bias removal uses two's complement to create a modified modulo-5 adder that removes a value $(011)_2 = 3$ from $|imp4| \dots |imp0\rangle$. Then, the result exponent sub-register $|ier2|ier1|ier0\rangle$ is prepared for the subsequent modulo-3 addition (denoted by *MADD*3) by applying *QFT*3. Next, the modulo-3 adder is used to add the qubits $|imp2|imp1|imp0\rangle$ into $|ier2|ier1|ier0\rangle$. By applying the inverse QFT3 on $|ier2|ier1|ier0\rangle$ the required state is obtained. The remaining steps shown in the quantum circuit in **Figure 3** are used to 'uncompute' and clean-up the temporary register, e.g. using inverse *QFT*3 and a modified modulo-5 adder to re-apply the bias $(011)_2 = 3$. The circuits described so far do not take into account the special situation arising from creating sub-normal numbers as output as well as cases with 'overflow' results. This is discussed next.

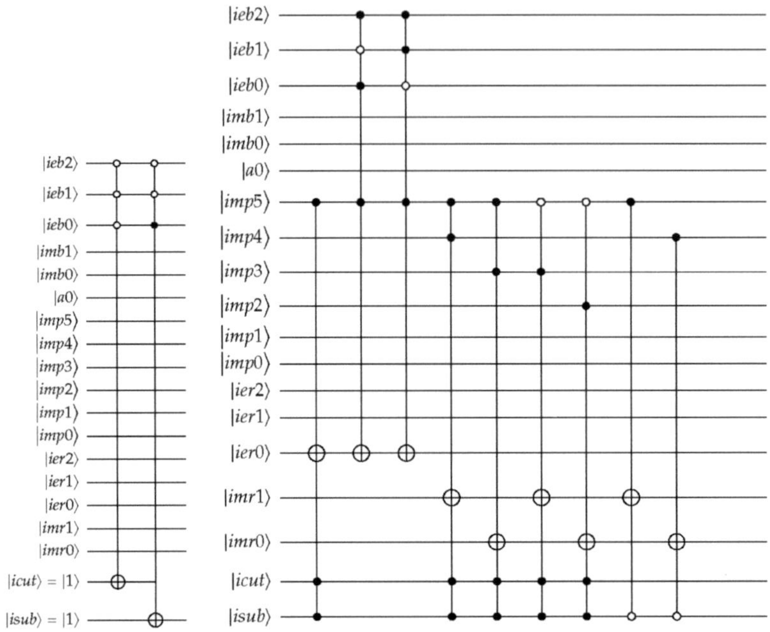

Figure 4.
Quantum circuits used in obtaining output mantissa for squaring operation, including sub-normal numbers and underflow/overflow protection ($N_M = 3$ and $N_E = 3$).

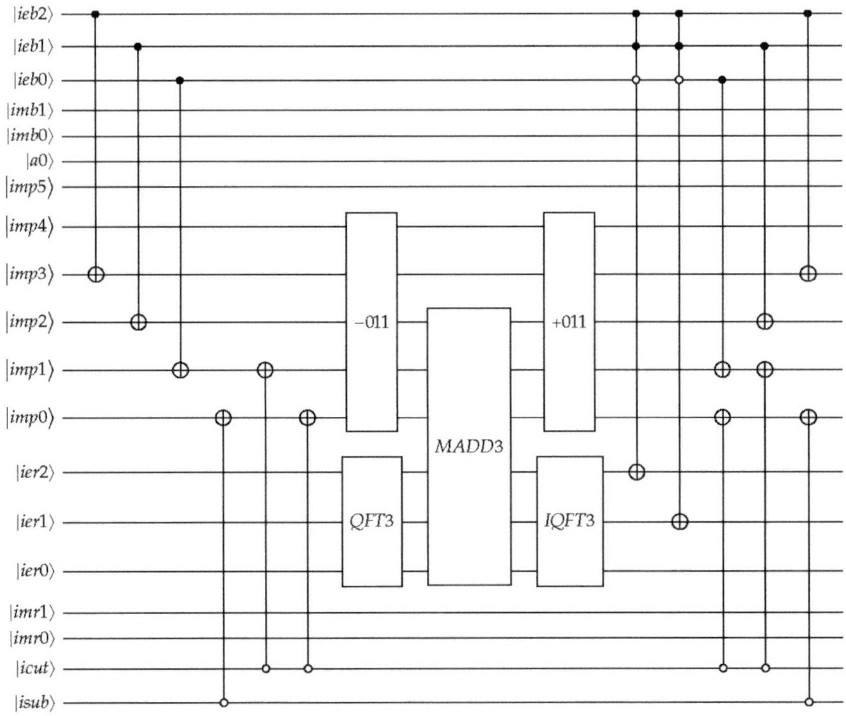

Figure 5.
Quantum circuit used to obtain exponent for squaring operation, including sub-normal numbers and under/overflow protection ($N_M = 3$ and $N_E = 3$).

For certain normalized input numbers the squaring operation leads to outputs truncated to 0 or to the non-zero sub-normal numbers discussed in Section 5.3. The quantum circuits discussed so far need to be modified in a number of ways to deal with this possible sub-normal output. **Figure 4** illustrates the required changes for $N_M = 3$ and $N_E = 3$. Two additional qubits are needed. Qubit $|isub\rangle = |0\rangle$ is used as indication that result is a sub-normal number. Qubit $|icut\rangle = |0\rangle$ is similarly used to define cases with output truncated to 0. Both qubits are initialized to $|1\rangle$. Then, before the mantissa multiplication step takes place, a first modification is introduced, shown on the left-hand side of **Figure 4**. For $N_E = 3$, only inputs with exponent $|000\rangle$ will need truncating to 0, as shown in the first 4-qubit controlled-NOT gate flipping $|icut\rangle$ to $|0\rangle$. For $N_E = 3$, inputs with exponent $|001\rangle$ are guaranteed to lead to sub-normal output (or 0), and for these cases $|isub\rangle$ is set to $|0\rangle$, using the second 4-qubit controlled-NOT gate with $|isub\rangle$ as target. The mantissa-multiplication step shown in **Figure 2** remains unchanged (i.e. qubits $|isub\rangle$ and $|icut\rangle$ are not used). The next required modification relates to the 'copying' of the result of the mantissa multiplication to output register $|imr1|imr0\rangle$ and the application of increments to the output exponent. The additional logic needed is

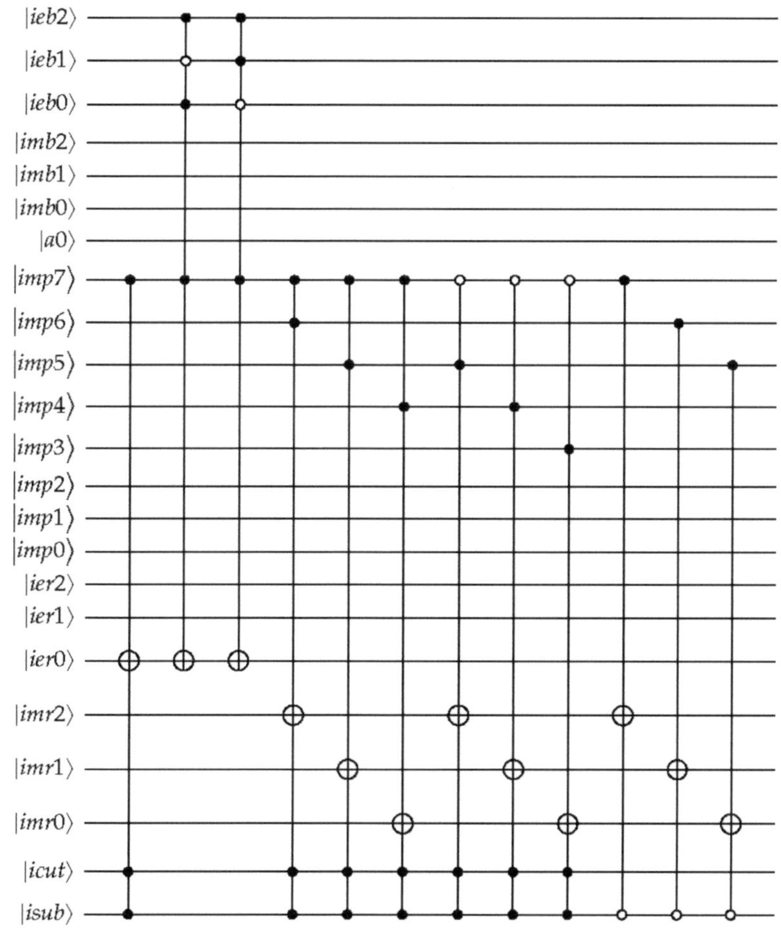

Figure 6.
Quantum circuit used to set output mantissa for squaring operation, including sub-normal numbers and underflow/overflow protection ($N_M = 4$ and $N_E = 3$).

shown on the right-hand side of **Figure 3**. First, for $|imp5\rangle = |1\rangle$, setting $|ier0\rangle = |1\rangle$ becomes conditional of both $|isub\rangle = |1\rangle$ and $|icut\rangle = |1\rangle$. The next two 4-qubit gates are used to guarantee that correct output with exponent $|111\rangle$ is created for inputs with exponents $|101\rangle$ and $|110\rangle$. The remaining gate operations perform the 'copying' of the mantissa squared into $|imr1|imr0\rangle$ taking into account the possible sub-normal output (cases with $|isub\rangle = |0\rangle$). The steps for $|isub\rangle = |1\rangle$ are the same as in the corresponding circuit for squaring without the sub-normal number modifications. A further set of circuit modifications to deal with sub-normal numbers is required in the quantum circuit used to obtain the output exponent. **Figure 5** shows the additional operations required relative to the original quantum circuit shown in **Figure 3**. Three additional CNOT operations are introduced just before performing the *QFT*3. For $|isub\rangle = |0\rangle$ and $|icut\rangle = |0\rangle$ the initialization of $|imp1|imp0\rangle$ is modified so that the subsequent steps will produce the correct result for the exponent. The three CNOT operations also appear in the 'uncompute' stage at the right-hand side of the circuit. Further changes comprise two 4-qubit controlled-NOT operations on $|ier2\rangle$ and $|ier1\rangle$ required to create $|111\rangle$ exponents for inputs with exponent $|110\rangle$.

For a fixed value of N_E it is important to note that the additional complexity introduced by increasing N_M is limited. In fact, the quantum circuit shown on the left-hand side of **Figure 4** does not depend on N_M. Similarly, the quantum circuits used to obtain the result exponent are independent of N_M. The circuit shown on the right-hand side of **Figure 4**, representing the definition of $|imr1|imr0\rangle$ for cases with normalized or sub-normal output requires modification. **Figure 6** shows how $|imr2|imr1|imr0\rangle$ are set for $N_M = 4$ using a set of gate operations that has grown

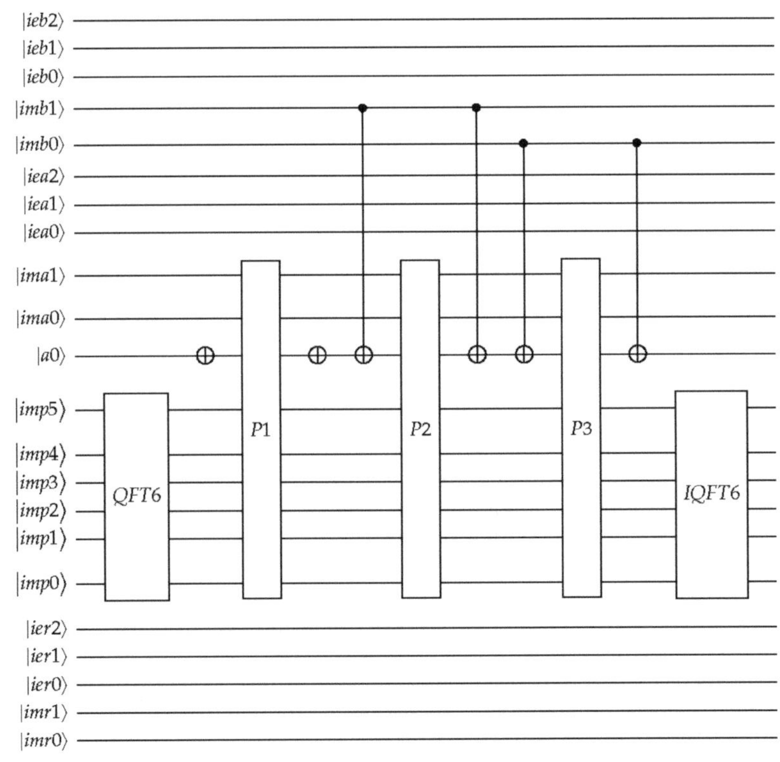

Figure 7.
Quantum circuit used in multiplying the mantissa of two input numbers ($N_M = 3$ and $N_E = 3$).

linearly with N_M. The circuit shown accounts for sub-normal numbers and includes underflow/overflow protection.

7. Quantum circuits for multiplication of floating-point numbers

In the interest of brevity, only the main features of the quantum circuits used for multiplication of two quantum floating-point numbers are summarized here. **Figure 7** illustrates the quantum circuit used to compute the product of the mantissas of two inputs. Compared to the circuit shown in **Figure 2** the main difference is that ancilla qubit $|a0\rangle$ is now set using the mantissa of a second input. A further difference relative to the squaring operation occurs in the circuit used to obtain the result exponent. Here, instead of setting $2\times$ the exponent using a bit shift, the sum of the two input exponents needs to be computed employing a quantum full adder.

8. Results of simulation and verification of quantum circuits

The proposed quantum circuits for squaring and multiplying floating-point numbers as part of the computational-basis representation, were systematically verified by gate-level simulation of the circuits for a wide range of cases with and without sub-normal numbers as well as cases with overflow results. The C++ quantum computer simulator detailed in previous work [4] was used for this purpose. To illustrate the process, the quantum algorithm used to square numbers with $N_M = 3$ and $N_E = 3$ is considered, with the following 19-qubit register (algorithm demonstrated accounts for sub-normal numbers as well as underflow/overflow protection, see Eq. (8) for reference):

$$|ieb2|ieb1|ieb0|imb1|imb0|a0|imp5|\ldots|imp0|ier2|ier1|ier0|imr1|imr0|icut|isub\rangle \quad (9)$$

where $|ieb2|ieb1|ieb0\rangle$ and $|imb1|imb0\rangle$ define the exponent and the fractional part of the mantissa of the input, respectively. Qubits $|icut\rangle$ and $|isub\rangle$ are initialized as $|1\rangle$, while all other qubits are initialized as $|0\rangle$. The quantum state in the simulation is then initialized with a single non-zero (unit) amplitude, with the index in the quantum state vector defined by the binary representation of input exponent and fractional part of mantissa. With the rounding mode fixed at rounding down to nearest, the intended output can be easily computed before the quantum circuit is simulated. In effect, this defines the index of the single non-zero (unit) amplitude of the output quantum state that should be returned in case the circuit is correct. Upon finalizing the quantum computer simulation the actual quantum state vector obtained is compared against the previously-computed required output. For this verification to be meaningful, the following range of possible inputs and outputs

Input	Initial state	Output state
7/2 (i)	$\psi_{init}\left[(1001100000000000011)_2\right] = 1$	$\psi_{out}\left[(1001100000001101011)_2\right] = 1$
7/16 (ii)	$\psi_{init}\left[(0011100000000000011)_2\right] = 1$	$\psi_{out}\left[(0011100000000001110)_2\right] = 1$
3/16 (iii)	$\psi_{init}\left[(0001100000000000011)_2\right] = 1$	$\psi_{out}\left[(0001100000000000001)_2\right] = 1$
6 (iv)	$\psi_{init}\left[(1011000000000000011)_2\right] = 1$	$\psi_{out}\left[(1011000000001110011)_2\right] = 1$

Table 3.
Results from quantum circuit simulation for representative range of inputs (squaring $N_M = 3$, $N_E = 3$).

were considered: (i) input and output are both normalized numbers, (ii) input is normalized number and output is a sub-normal number, (iii) input is a sub-normal number and result truncated to 0, (iv) input is a normalized number, with output overflow. For $N_M = 3$ and $N_E = 3$, **Table 3** summarizes the input and output states for examples of each of the 4 categories considered. For inital and output the single non-zero amplitudes are shown. Since the simulator employed here stores the full 2^{n_q} state vector for n_q qubits, only circuits with ≤ 28 qubits were considered as a result of limited computational resources and the large number of cases considered (> 100). For the squaring operation, $N_M \in [3, 6]$ and $N_E \in [3, 4]$ were considered, while for multiplication the range of N_M needed to be reduced, i.e. $N_M \in [3, 4]$.

N_M	N_E	$L_2(u)$	$L_\infty(u)$	$L_2(p)$	$L_\infty(p)$
Rounding down - using sub-normal numbers					
3	3	26.805	0.124741	13.2908	0.0623342
4	3	7.69964	0.0624983	3.79396	0.0310842
5	3	1.93069	0.0312483	0.883095	0.0154592
6	3	0.477862	0.0156233	0.233542	0.00780768
7	3	0.110358	0.00781078	0.0611784	0.00390143
8	3	0.0247615	0.00390453	0.0135501	0.00194831
4	4	6.36002	0.0624983	1.57508	0.0310387
5	4	1.62679	0.0312483	0.387261	0.0154137
6	4	0.409663	0.0156233	0.10847	0.00762945
7	4	0.0958982	0.00781078	0.0296086	0.0037232
8	4	0.0209894	0.00390453	0.00647854	0.00192175
Rounding down - without sub-normal numbers					
3	3	86.8625	0.248583	111.896	0.249507
4	3	70.4413	0.248583	108.352	0.249507
5	3	65.8235	0.248583	107.359	0.249507
6	3	64.6349	0.248583	107.135	0.249507
7	3	64.3262	0.248583	107.069	0.249507
8	3	64.2529	0.248583	107.050	0.249507
4	4	6.3881	0.0624983	1.6114	0.0310387
5	4	1.65503	0.0312483	0.42405	0.0154137
6	4	0.437976	0.0156233	0.14534	0.0151248
7	4	0.124223	0.0147218	0.0665074	0.0151248
8	4	0.0493163	0.0147218	0.0433827	0.0151248

Table 4.
Approximation errors in Taylor-green vortex flow field due to reduced-precision floating-point representation. L_∞ and L_2 norms of errors relative to IEEE double-precision representation for velocity (u) and pressure (p) for different N_M and N_E. 100×100 uniform mesh.

9. Complexity analysis

Before analyzing the quantum circuits introduced here in terms of complexity, first the choice of N_M and N_E for representing realistic flow fields is considered.

9.1 Representing Taylor-green vortex flow

In a two-dimensional flow field, the non-linear terms appearing in the Navier–Stokes equations, shown in Eq. (4), involve the square of the velocity components in $x-$ and $-y$ directions, i.e. u^2 and v^2, as well as, the product uv. Here, the example flow field defined by the two-dimensional Taylor-Green vortex is considered, where velocity and pressure are defined in a square domain $[0, 2\pi]^2$ with periodic boundary conditions as,

$$u = cos\,(x)\,sin\,(y) \;\; ; \quad v = -\,sin\,(x)\,cos\,(y) \;\; ; \quad p = -\frac{1}{4}[cos\,(2x) + cos\,(2y)] \tag{10}$$

Considering a 100×100 uniform mesh, the effect of representing the flow field variables with a reduced-precision floating-point format is analyzed first.

N_M	N_E	$L_2(u^2)$	$L_\infty(u^2)$	$L_2(\lvert uv \rvert)$	$L_\infty(\lvert uv \rvert)$
Rounding down - using sub-normal numbers					
4	4	0.801596	0.0351562	0.161925	0.0146484
5	4	0.3848	0.0244141	0.0520772	0.00732422
6	4	0.101035	0.013916	0.018016	0.00378418
7	4	0.0382158	0.00738525	0.0053449	0.00186157
8	4	0.0108621	0.00379944	0.00123537	0.000919342
Rounding down - without sub-normal numbers					
4	4	0.87222	0.0351562	0.30511	0.0147705
5	4	0.461689	0.0244141	0.213756	0.0153809
6	4	0.18035	0.0151405	0.188371	0.0154495
7	4	0.119222	0.0151405	0.179671	0.0154495
8	4	0.0927176	0.0152609	0.177551	0.015553

Table 5.
Approximation errors of velocity products in Taylor-green vortex flow field due to reduced-precision floating-point representation. L_∞ and L_2 norms of errors relative to IEEE double-precision representation for velocity (u^2) and pressure (uv) for different N_M and N_E. 100×100 uniform mesh.

	CPHASE	C^2PHASE	θ_{min}
3×3	9	27	$2\pi/2^6$
4×4	14	66	$2\pi/2^8$
5×5	20	130	$2\pi/2^{10}$

Table 6.
Number of controlled-phase gates (CPHASE) and doubly-controlled-phase (C^2PHASE) for phase-addition operator in quantum-multiplier. Also, smallest rotation angle is shown.

	CPHASE	θ_{min}		CPHASE	θ_{min}
MADD3	6	$2\pi/2^3$	*FADD3*	9	$2\pi/2^4$
MADD4	10	$2\pi/2^4$	*FADD4*	14	$2\pi/2^5$
MADD5	15	$2\pi/2^5$	*FADD5*	20	$2\pi/2^6$
MADD6	21	$2\pi/2^6$			

Table 7.
Number of controlled-phase gates (CPHASE) in phase-addition step for modulo adder (MADD) and full adder (FADD). Also, smallest rotation angle is shown.

Table 4 summarizes the results, highlighting the importance of including sub-normal numbers in the floating-point representation. Since a sign bit is not used here, the absolute values of u, v, p were actually used. Flow variables defined in Eq. (10) are in the range $[-1, 1]$, so that by increasing N_E from 3 to 4, far fewer sub-normal numbers are used to represent the flow field. As a result, removing the sub-normal number capability (as shown in bottom half of table), results in smaller errors for $N_E = 4$. For realistic applications of the proposed quantum floating point format, the relatively small overhead incurred by introducing sub-normal numbers in the quantum circuits clearly suggests that sub-normal numbers should be included.

For $N_E = 4$, the representation of u^2 and $|uv|$ is considered. Specifically, the error shown is that introduced by the multiplication: the difference between the 'exact' product of the reduced-precision representation of $|u|$ and $|v|$ and the corresponding reduced precision representation of the products is shown in **Table 5**. The results highlight that although sub-normal numbers played a relatively smaller role in representing velocity components, in the computation of the nonlinear terms, the inclusion of sub-normal numbers is more important for the minimization of approximation errors.

9.2 Mantissa multiplication step

QFT and inverse *QFT* are used involving $2N_M$ qubits, so that the complexity in terms of two-qubit (controlled-phase) gates scales as N_M^2, where the well-known complexity of the standard *QFT* implementation is used. The complexity of the phase-addition steps involved in the multiplication are detailed in **Table 6**. For the two-qubit gates the number can be seen to scale as N_M^2, while the number of three-qubit gates shows a N_M^3 scaling.

9.3 Computation of exponent

QFT and inverse *QFT* are used involving N_E, $N_E + 1$ and $N_E + 2$ qubits, representing a smaller complexity than the *QFT* used in mantissa multiplications. The main contributions to complexity of exponent computation stems from the modulo and full-adders involving a number of qubits scaling linearly with N_E. The polynomial complexity in terms of qubits for the adders implemented here is shown in **Table** 7.

9.4 Discussion

The quantum circuits presented here for squaring two floating-point numbers in the format proposed show that by accounting for sub-normal numbers and under/overflow an additional number of multi-qubit controlled-NOT gates is needed. However, for the examples analyzed a polynomial dependence on N_M and N_E was observed. This means that in terms of quantum-algorithm complexity this implementation has the desired efficiency. The relatively small complexity as compared to circuits used for mantissa multiplication highlights that for most applications it is desirable to include the capability of using sub-normal numbers and provide under/overflow protection in the quantum circuits. The analysis in this section also shows that for a realistic application, a well-considered scaling of the governing equations to $O(1)$ variables is even more important here than in classical implementations using IEEE single- or double-precision arithmetic. Using the limited number of qubits available on current and near-term quantum computers (<100), the proposed approach to introducing non-linearity is a good candidate in cases where N_M and N_E can be chosen significantly smaller than in equivalent classical floating-point representations.

10. Conclusions

The challenges associated with representing non-linear differential equations in terms of quantum circuits were discussed in this chapter. In this work, a new approach for representing product-terms in nonlinear equations suitable for near-term (e.g. NISQ generation) quantum computers was proposed. A key aspect discussed is the (temporary) representation of the variables in the computational basis. Furthermore, the use of a suitably-chosen floating-point format was detailed. The importance of including sub-normal numbers, such as defined in the IEEE 758 standard for floating-point arithmetic on classical computers, was demonstrated. Based on the current findings, a number of suggestions for further work can be put forward. The presented circuits performed arithmetic for a single set of input data, i.e. equivalent to data for a single point in a computational domain. Extending the approach to a multi-dimensional computational mesh is a first step to consider. A complexity analysis will be needed to assess the potential speed-up relative to classical discretization approaches for the considered equations. A further step involves investigating how the proposed approach can be made part of a larger quantum algorithm, where a mix of amplitude-based encoding and computational-basis encoding occurs. A key aspect is therefore the development of efficient quantum circuits to perform the required conversions between the two different encoding approaches. Finally, further work is needed to establish how the approach presented here can be used in a wider range of quantum computing applications.

Author details

Rene Steijl
James Watt School of Engineering, University of Glasgow, Glasgow, United Kingdom

*Address all correspondence to: rene.steijl@glasgow.ac.uk

References

[1] Nielsen MA, Chuang IL. Quantum Computation and Quantum Information: 10th Anniversary Edition. 2nd ed. Cambridge: Cambridge University Press; 2010.

[2] Yepez J. Quantum lattice-gas model for computational fluid dynamics. Phys. Rev. E. 2001;63(4):046702. DOI: 10.1103/PhysRevE.63.046702

[3] Berman GP, Ezhov AA, Kamenov DI, Yepez J. Simulation of the diffusion equation on a type-II quantum computer. Phys. Rev. A. 2002;66(1): 012310. DOI:10.1103/PhysRevA. 66.012310

[4] Steijl R, Barakos GN. Parallel evaluation of quantum algorithms for computational fluid dynamics. Computers&Fluids 2018;173:22-28. DOI: 10.1016/j.compfluid.2018.03.080

[5] Steijl R. Quantum Algorithms for Fluid Simulations. In: Bulnes F, Stavrou VN, Morozov O, Bourdine AV, editors. Advances in Quantum Communication and Information. IntechOpen; 2020. DOI: 10.5772/ intechopen.86685

[6] Todorova BN, Steijl R. Quantum Algorithm for the collisionless Boltzmann equation. J. Comp. Phys. 2020;409:109347. DOI:10.1016/j. jcp.2020.109347

[7] Griffin KP, Jain SS, Flint TJ, Chan WHR. Investigations of quantum algorithms for direct numerical simulation of the Navier-Stokes equations. Center for Turbulence Research Annual Research Briefs. 2019; 347-363.

[8] Gaitan F. Finding flows of a Navier–Stokes fluid through quantum computing. npj Quantum Information. 2020; 6:61. DOI:10.1038/s41534-020-00291-0

[9] Harrow AW, Hassidim A, Lloyd S. Quantum algorithm for linear systems of equations. Phys. Rev. Lett. 2009;103 (15):150502. DOI:10.1103/ PhysRevLett.103.150502

[10] Clader BD, Jacobs BC, Sprouse CR, Preconditioned quantum linear system algorithm. Phys. Rev. Lett. 2013;110 (25):25054. DOI:10.1103/ PhysRevLett.110.250504

[11] Cao Y, Papageorgiou A, Petras I, Traub J, Kais S. Quantum algorithm and circuit design solving the Poisson equation. New J. Phys. 2013;15:013021. DOI:10.1088/1367-2630/15/1/013021

[12] Scherer A, Valiron B, Mau S-C, Alexander S, van den Berg E, Chapuran TE. Concrete resource analysis of the quantum linear-system algorithm used to compute the electromagnetic scattering cross section of a 2D target. Quantum Inf. Proc. 2017;16(3):60. DOI: 10.1007/s11128-016-1495-5

[13] Montanaro A, Pallister S. Quantum Algorithms and the finite element method. Phys. Rev. A. 2016;93(3): 032324. DOI:10.1103/ PhysRevA.93.032324

[14] Xu G, Daley AJ, Givi P, Somma RD. Turbulent mixing simulation via a quantum algorithm. AIAA J. 2018;56(2): 687-699. DOI:10.2514/1.J055896

[15] Berry DW. High-order quantum algorithm for solving linear differential equations. J. Phys. A. 2014;47(10): 105301. DOI:10.1088/1751-8113/47/10/ 105301

[16] Berry DW, Childs AM, Ostrander A, Wang G. Quantum Algorithm for Linear Differential Equations with Exponentially Improved Dependence on Precision. Comm. Math. Phys. 2017;356 (3):1057-1081. DOI:10.1007/ s00220-017-3002-y

[17] Fillion-Gourdeau F, Lorin E. Simple digital quantum algorithm for symmetric first-order linear hyperbolic systems. Numerical Algorithms. 2019; 82:1009-1045. DOI:10.1007/s11075-018-0639-3

[18] Costa PCS, Jordan S, Ostrander *A. quantum* algorithm for simulating the wave equation. Phys. Rev. A. 2019;99 (1):012323. DOI:10.1103/ PhysRevA.99.012323

[19] Childs AM, Liu J-P. Quantum spectral methods for differential equations. Comm. Math. Phys. 2020; 375(2):1427-1457. DOI:10.1007/ s00220-020-03699-z

[20] Leyton SK, Osborne TJ. A quantum algorithm to solve nonlinear differential equations. arXiv.org 2008;0812.4423.

[21] Zhou SS, Loke T, Izaac JA, Wang JB. Quantum Fourier transform in computational basis. Quantum Inf. Proc. 2017; 16(3):82. DOI:10.1007/ s11128-017-1515-0

[22] Peruzzo A, McClean J, Shadbolt P, Yung M-H, Zhou X-Q, Love PJ, Aspuru-Guzil A, O'Brien JL. A variational eigenvalue solver on a photonic quantum processor. Nature Comms 2014; 5:4213. DOI:10.1038/ncomms5213

[23] McClean JR, Romero J, Babbush R, Aspuru-Guzik A. The theory of variational hybrid quantum-classical algorithms. New J. Phys. 2016; 18:023023. DOI:10.1088/1367-2630/18/2/023023

[24] Lubasch M, Joo J, Moinier P, Kiffner M, Jaksch D. Variational quantum algorithms for nonlinear problems. Phys. Rev. A. 2020;101(1): 010301. DOI:10.1103/ PhysRevA.101.010301

[25] Mitarai K, Kitagawa M, Fijii K. Quantum analog-digital conversion. Phys. Rev. A. 2019;99(1):012301. DOI: 10.1103/PhysRevA.99.012301

[26] SaiToh *A. quantum* digital-to-analog conversion algorithm using decoherence. Quantum Inf. Proc. 2015; 14(8):2729-2748. DOI:10.1007/ s11128-015-1033-x

[27] Haener T, Soeken M, Roetteler M, Svore KM. Quantum circuits for floating-point arithmetic. In: Kari J, Ulidowski I, editors. Reversible Computation. RC 2018. Lecture Notes in Computer Science, vol 11106. Springer; 2018. DOI:doi.org/10.1007/978-3-319-99498-7-11

[28] Grover L, Rudolph T. Creating superpositions that correspond to efficiently integrable probability distribution. arXiv 2002;0208112

[29] Ma G, Li H, Zhao J. Quantum QR decomposition in the computational basis. Quantum Inf. Proc. 2019; 19:271. DOI:10.1007/s11128-020-2777-4

[30] Bhaskar MK, Hadfield S, Papageorgiou A, Petras I. Quantum algorithms and circuits for scientific computing. Quantum Info. Comput. 2016;16(3-4):197-236. DOI:10.5555/ 3179448.3179450

[31] Overton M.L. Numerical Computing with IEEE Floating Point Arithmetic. 1st ed. Philadelphia: SIAM; 2001. 97p.

Chapter 3

A Novel Three-Input XOR Gate Based on Quantum Dot-Cellular Automata with Power Dissipation Analysis

Ismail Gassoumi, Lamjed Touil and Abdellatif Mtibaa

Abstract

Recently, Low power and reduced heat dissipation are an increasing demand for digital systems. Quantum Dot Cellular Automata (QCA) is a future generation solution based on nanotechnology for the digital systems. The QCA systems have advantages like the small size, ultralow power consumption and high switching frequency. The present research aims at introducing a novel three-input XOR gate containing 12 cells. The energy dissipation analysis of the proposed gate is verified using three different energy levels ($\gamma = 0.5\,E_k$, $\gamma = 1.0$, E_k and $\gamma = 1.5\,E_\kappa$) at T = 2 Kelvin temperature. Simulation is performed for the proposed gate using QCA Designer tool version 2.0.3. The proposed three-input XOR gate has less number of cells, area and energy dissipation as compared to the previous structures.

Keywords: nanotechnology, circuit design, quantum-dot cellular automata (QCA), three-input XOR gate

1. Introduction

The state of the art very large scale integrated circuits (VLSI) technology limits to doping fluctuations and high leakage current [1]. On one side, scaling down of CMOS technology has led to grave challenges in context of power consumption, physical dimensions, and leakage current. These short falls have guided to significant efforts to look for suitable substitutes. On the other side, emerging nanotechnologies seems to be better choice for the future generation digital systems [2, 3]. Thereby, quantum computers promise dramatic improvements in our ability to efficiently solve classically intractable problems ranging from cryptosystems to simulation of quantum systems. Quantum computing has attracted attention in the past two decades because it was found that computers exploiting quantum mechanics are able to outperform classical digital computers in certain areas like factoring integers and searching. Developments in the field of quantum computing have been strongly impacted by the paradigm of quantum-dot cellular automata (QCA), in which information is transmitted and processed through electrostatic interactions in an array of cells. QCA is one of the most significant computing technologies for the future. It will be the alternative candidate for CMOS technology that currently used in integrated circuits (ICs) [2, 3]. The logic function of QCA technology is to implement circuits using movement of electrons rather than voltage level [3–6]. In fact, QCA technology is anticipated to offer higher density and lower power consumption and more flexible interconnection designs for future System on Chip (SoC). On the other hand, a number of QCA based digital devices have studied to date; designs of XOR gate, full adder, multipliers, dividers, memory circuits, counter QCA based memory cells, flip flops, and multiplexer [7–15]. Among theme, XOR gates are extensively employed in communication systems. So, there is emerging need to develop methods which involve less area and delay overheads to improve the complexity of digital circuits. With this motive, we have proposed a novel design of QCA based three-input XOR gate. The proposed gate has significant improvement compared to others design prestened in the literature.

2. Basics of QCA

No voltage or current is used. It is possible to replace the gate of a transistor by a molecular charge center and encode information in its charge state. The electrons residing in the diagonally opposite positions lead to two equivalent energy states representing logic '0' and logic '1' which are respectively called as cell polarizations P = +1.00 and P = −1.00 as shown in **Figure 1b**. The clocking is the key element of QCA circuitry as shown in **Figure 1a**. The two basic logic gates in QCA are inverter and majority voter as depicted in **Figure 1d** and **e**. A cell changes its polarization based on the fixed polarization of the cell placed by its side. This feature of QCA cell is exploited when QCA cells arranged in a series act similar to wire as shown in **Figure 1c**.

3. Related works

3.1 An overview of previous 3-input QCA XOR gates

Exclusive OR (XOR) is an applicable gate for designing the most of logic circuits. This gate has a wide range of applications, particularly in designing circuits such as full-adders, multipliers, dividers, and compressors. Until now, a number of XOR gates have been reported. Angizi et al. in [16] reported an XOR gate that required 94 cells and occupied 0.073 μm^2 area. Moreover, the time delay of this gate is 1.5 clock cycles. One of the simplified structures of the XOR gate has been introduced by Ahmad et al., in [17], which requires 14 cells, occupies 0.022 μm^2 area and output appears after 0.5 clock cycle. However, this gate cannot achieve the expected

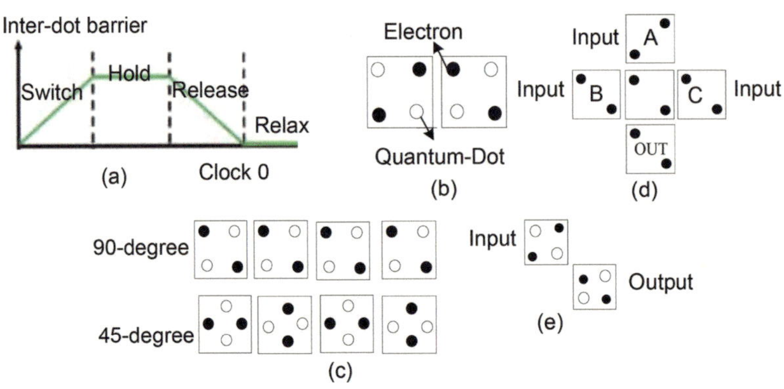

Figure 1.
(a) Four stages of clock (b) logic "0" and "1" states (c) wires for circuit (d) 3-input MG (e) invert.

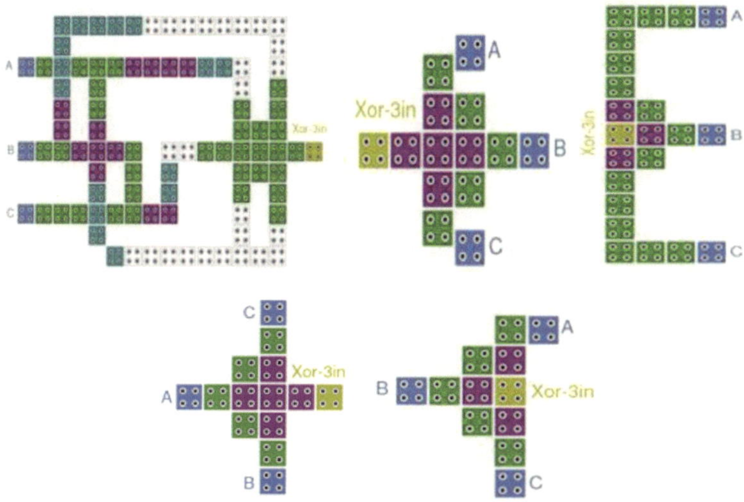

Figure 2.
The QCA representation of the previous 3-input XOR gate in (a) [20], (b) [19], (c) [17], (d) [18], (e) [16].

optimization for the larger circuit. Bahar et al. have reported another compact XOR gate, in [18] that used 12 cells and occupied 0.012 μm^2 area. However, this gate is not suitable for designing large scale circuits. Balali et al., in [19], proposed another 14 cells XOR gate; however, the use of half-cell translation inverter gates make this gate more impractical in terms of physical realization. More recently, Bahar et al., in [20], claimed that the proposed E-shaped XOR gate is capable of achieving higher designing optimization at a more extensive design paradigm. In the following, a unique ultra-efficient XOR gate is proposed. The QCA layout of this gate is simple, efficient and appropriate to design of all logical functions. In addition, **Figure 2** depicts various layouts of previously 3-input XOR gate in QCA presented in the literature.

4. The proposed three-input QCA XOR gate

Exclusive-OR (XOR) is the most fundamental component used in digital circuits including parity generator and checker, comparator, code converter, arithmetic and logic processing unit, and so on. QCA layout of the proposed

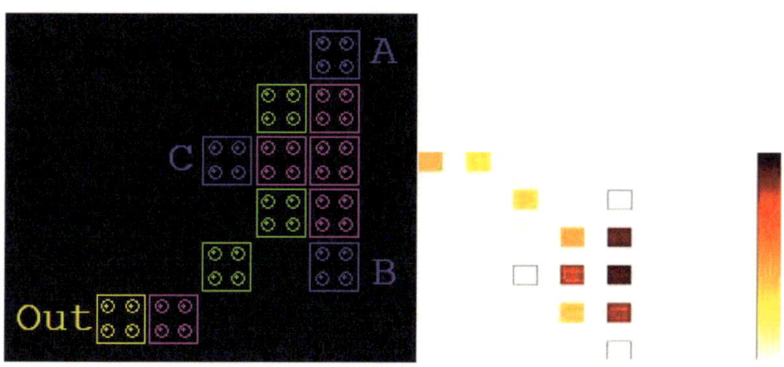

(a) *(b)*

Figure 3.
The QCA layout of the proposed three-input XOR gate and its power dissipation map.

three-input XOR gate and its power dissipation map are shown in **Figure 3(a)** and **(b)** respectivly, which consists 12 cells with occupied area is 0.01 μm^2 and requires two clock phases to generate the corrects output. It is clear in the suggested layout that there is no majority gate, resulting in reduced space and energy consumed. In fact, the presented layout utilize electrostatic interactions between cells within QCA configurations to perform desired function.

5. Results and discussions

Results of the simulation of the suggested QCA-gate is presented in this section obtained using the CAD tool QCADesigner. Coherence vector simulation engine and all other parameters set at default values are used. The essential QCA parameters are presented in **Table 1**. Reduction in the XOR gate size will result in a subsequent reduction of the scaled-up circuits. The simulation result of the proposed three input

Parameter	Value
Number of samples	12800
Convergence tolerance	0.001000
Radius of effect	65,000000(nm)
Relative permittivity	12,900000
Clock low	3,800000e-023
Clock high	9,800000e-022
Clock shift	0,000000e+000
Clock amplitude factor	2,000000
Layer separation	11,500000
Maximum iterations per sample	100

Table 1.
Bistable approximation parameters model.

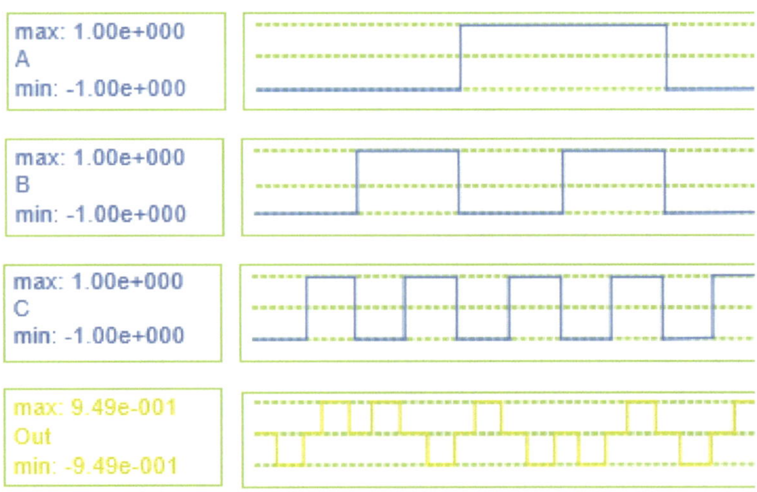

Figure 4.
Output waveforms of the proposed gate.

Three-input XOR	Area(μm2)	Cell count	Latency
Ref [18]	0.012	12	1
Ref [16]	0.017	22	1
Ref [20]	0.073	94	1.5
Ref [19]	0.022	14	0.5
Ref [17]	0.011	14	0.5
Proposed Design	0.010	12	0.5

Table 2.
The comparison of the 3-input QCA XOR gates.

Three-input XOR gate	Total energy dissipation		
	0.5 EK	1 EK	1.5 EK
Ref [18]	47.29	62.39	80.34
Ref [16]	146.44	171.57	204.47
Ref [20]	36.20	50.28	66.58
Ref [20]	49.81	63.49	80.83
Ref [17]	12.11	14.17	16.28
Proposed XOR	10.43	12.25	14.32

Table 3.
Power analysis results of the proposed 3-input XOR gate and previously reported designs.

XOR gate is shown in **Figure 4**. The comparison outcomes of the number of consumed cells (cell count), the occupied area, the gate count, and the latency of the suggested 3-input XOR with the previous coplanar 3-input XOR gate in Refs [16, 21, 22] are shown in **Table 2**. The proposed QCA XOR gate represents a 14.28% improvement in cell consumption relative to the best-optimized gate in Ref. [16]. To estimate the power dissipation, we use the QCAPro software. **Table 3** depicts the detailed power dissipation data of the proposed QCA XOR gate at a temperature of 2 K. As expected, the proposed XOR gate dissipates 71.18%, 75.63%, and 78.49% less energy at 0.5 Ek, 1 Ek, and 1.5 Ek, respectively, compared with the XOR gate in [20]. It can be seen from **Table 3** that the proposed gate consumes the lowest amount of energy over previous designs, and therefore it is very appropriate for ultralow power devices.

6. Conclusions

The development of nano-scale quantum dot cellular automata (QCA) has been drivern by the immense need for high performance and energy-efficient computational systems. The present work proposed a new three-input QCA XOR gate consisting of 12 cells with an occupied space of 0.01 μm^2. Hence, the proposed gate is superior than the existing XOR structures in literature. The designed gate dissipate less energy, has been verified using the QCAPro tool. Simulation results have been shown that the suggested gate is suitable techniques to implement efficient logic circuits for QCA.. In the future, we will strive to explore and construct more excellent QCA based-design in order to provide basic module for the larger scale arithmetic operation circuits.

Author details

Ismail Gassoumi*, Lamjed Touil and Abdellatif Mtibaa
Laboratory of Electronics and Microelectronics, University of Monastir, Monastir, Tunisia

*Address all correspondence to: gassoumiismail@gmail.com

References

[1] C. S. Lent, P. D. Tougaw, W. Porod, G. H. Bernstein, "Quantum cellular automata," Nanotechnology, 4, 49-57, 1993.

[2] Smith, C. G., "Computation without current", Science, "284(5412), 274, 1999.

[3] P. D. Tougaw and C.S. Lent. "Logical devices implemented using quantum cellular automata," J. Appl. Phys. 75, 1818, 1994

[4] Frost, S. E., Rodrigues, A. F., Janiszewski, A. W., Rausch, R. T., & Kogge, P. M., "Memory in motion: A study of storage structures in QCA," In First Workshop on Non-Silicon Computing, vol. 2, Feb. 2002.

[5] Niemier, M. T., & Kogge, P. M., "Logic in wire: using quantum dots to implement a microprocessor," In Electronics, Circuits and Systems, Proceedings of ICECS'99. The 6th IEEE International Conference on, vol. 3, pp. 1211-1215. 1999.

[6] Amlani, I., Orlov, A. O., Toth, G., Bernstein, G. H., Lent, C. S., & Snider, G. L., "Digital logic gate using quantum-dot cellular automata," Science, 284(5412), 289-291, 1999.

[7] McDermott, L.C., Research on conceptual understanding in mechanics. Physics Today, 1984. 37: 44. p. 24-32

[8] Barughi, Y.Z. and S.R. Heikalabad, A three-layer full adder/subtractor structure in quantum-dot cellular automata. International Journal of Theoretical Physics, 2017. 56(9): p. 2848-2858.

[9] Sasamal, T.N., A.K. Singh, and A. Mohan, An optimal design of full adder based on 5-input majority gate in coplanar quantum-dot cellular automata. Optik-International Journal for Light and Electron Optics, 2016. 127(20): p. 8576-8591.

[10] Mohammadi, M., M. Mohammadi, and S. Gorgin, An efficient design of full adder in quantum-dot cellular automata (QCA) technology. Microelectronics Journal, 2016. 50: p. 35-43.

[11] Labrado, C. and H. Thapliyal, Design of adder and subtractor circuits in majority logic-based field-coupled QCA nanocomputing. Electronics Letters, 2016. 52(6): p. 464-466.

[12] Hayati, M. and A. Rezaei, Design of novel efficient adder and subtractor for quantum-dot cellular automata. International Journal of Circuit Theory and Applications, 2015. 43(10): p. 1446-1454.

[13] Kianpour, M., R. Sabbaghi-Nadooshan, and K. Navi, A novel design of 8-bit adder/subtractor by quantum-dot cellular automata. Journal of Computer and System Sciences, 2014. 80(7): p. 1404-1414.

[14] Sen, B., A. Rajoria, and B.K. Sikdar, Design of efficient full adder in quantum-dot cellular automata. The Scientific World Journal, 2013. 2013.

[15] Navi, K., et al., A new quantum-dot cellular automata full-adder. Microelectronics Journal, 2010, 41(12): p. 820-826.

[16] Ahmad, F., et al., Towards single layer quantum-dot cellular automata adders based on explicit interaction of cells. Journal of Computational Science, 2016. 16: p. 8-15.

[17] Angizi, S., et al., Novel robust single layer wire crossing approach for exclusive or sum of products logic design with quantum-dot cellular automata. Journal of Low Power Electronics, 2014. 10(2): p. 259-271.

[18] Bahar, A.N., et al., A novel 3-input XOR function implementation in quantum dot-cellular automata with energy dissipation analysis. Alexandria Engineering Journal, 2017.

[19] Balali, M., et al., Towards coplanar quantum-dot cellular automata adders based on efficient three-input XOR gate. Results in Physics, 2017. 7: p. 1389-1395.

[20] Bahar, A.N. and K.A. Wahid, Design of QCA-Serial Parallel Multiplier (QSPM) with Energy Dissipation Analysis. IEEE Transactions on Circuits and Systems II: Express Briefs, 2019.

[21] Gladshtein, M., Design and simulation of novel adder/subtractors on quantum-dot cellular automata: Radical departure from Boolean logic circuits. Microelectronics Journal, 2013. 44(6):p. 545-552.

[22] Cho, H. and E.E. Swartzlander Jr, Adder and multiplier design in quantum-dot cellular automata. IEEE Transactions on Computers, 2009. 58(6): p. 721-727.

Chapter 4

Topology in Photonic Discrete-Time Quantum Walks: A Comprehensive Review

Graciana Puentes

Abstract

We present a comprehensive review of photonic implementations of discrete-time quantum walks (DTQW) in the spatial and temporal domains. Moreover, we introduce a novel scheme for DTQWs using transverse spatial modes of single photons and programmable spatial light modulators (SLM) to manipulate them. We discuss current applications of such photonic DTQW architectures in quantum simulation of topological effects in photonic systems.

Keywords: quantum walks, spatial-multiplexing, time-multiplexing, spatial light modulators, geometric phase, Zak phase, topology

1. Introduction

Quantum computation is an interdisciplinary field that encompasses several interconnected branches such as quantum algorithms, quantum information, and quantum communication. There are several advantages associated with quantum information processing that have positioned quantum computation as a key resource in advanced modern science and technologies. Among the promising conjectures predicted by quantum information and communication, we find the development of more powerful algorithms that may allow to significantly increase the processing capacity and may enable the quantum simulation of complex physical systems and mathematical problems for which we know no classical digital computer algorithm that could efficiently simulate them at present.

Quantum algorithms are the main building blocks of quantum information and quantum communication strategies. Nevertheless, building superior quantum algorithms is a challenging task due to the complexities of quantum mechanics itself, and because quantum algorithms are required to demonstrate that they can outperform their classical counterparts, in order to be considered an evolutionary advantage. Therefore quantum algorithms must be more efficient than any existing classical protocol. In this context, quantum walks, i.e., the quantum mechanical counterpart of classical random walks, can be regarded as a sophisticated tool for building quantum algorithms for quantum information and quantum communication that has been shown to constitute a universal model for quantum computation [1–13].

The quantum walk is one of the most striking manifestations of how quantum interference leads to a strong departure between quantum and classical phenomena [2, 3, 14]. In the discrete version of the quantum walk, namely the discrete-time quantum walk (DTQW) [15], the time evolution is described in terms of a series o discrete time-steps. DTQWs provide for a flexible architecture for the investigation of a large number of complex topological and holonomical effects, in the experimental [16–18] and theoretical domains [19–31]. Moreover, DTQWs are robust algorithm for modeling a large number of time-varying processes, ranging from energy transfer in chains of spins [32, 33] to energy transport in biological systems [34]. Furthermore, DTQWs allow to study multi-dimensional quantum interference effects [35–38] and can outline a route for authentication of quantum complexity [39, 40] and universal quantum computation [41]. In addition, quantum walks involving multiple particles guarantee a relentless tool for encoding quantum information in an exponentially large Hilbert space [42], as well as for simulations in quantum chemical, biological and physical systems [43], in 1D and 2D geometries [44–46].

In this Chapter, we present a comprehensive review of photonic realizations of DTQW in both, the spatial [47] and the temporal [48] realms, based on spatial-multiplexing and time-multiplexing techniques, respectively. Moreover, we present a novel scheme for photonic DTQW exploiting transverse spatial modes of photons and programmable spatial light modulators (SLM) to manipulate the modes [3]. In contrast to all previous multiplexed implementations, this novel approach warrants quantum simulation of arbitrary discrete time-steps, only limited by the spatial resolution of the SLM itself. We also deliberate about possible applications of such photonic DTQW platforms in quantum simulation of topological phenomena in photonic systems, and the implementation of non-local quantum coin operations, based on two-photon hybrid entanglement. Part of this review is based on the work by the Author, selected as the cover story of a Special Issue on Quantum Topology, for the journal Crystals (MDPI) in 2017 [2].

2. Theoretical framework

As a starter, we describe the theoretical framework for the mathematical description of DTQWs, and applications in the generation and detection of non-trivial geometric-phase structures, in 1D DTQW platforms. The basic discrete step in the DTQW is mathematically described by a unitary quantum evolution operator $U(\theta) = TR_{\vec{n}}(\theta)$, with $R_{\vec{n}}(\theta)$ a rotation operation along an arbitrary direction, represented by the 3D vector $\vec{n} = (n_x, n_y, n_z)$, represented by the following expressions:

$$R_{\vec{n}}(\theta) = \begin{pmatrix} \cos(\theta) - in_z \sin(\theta) & (in_x - n_y)\sin(\theta) \\ (in_x + n_y)\sin(\theta) & \cos(\theta) + in_z \sin(\theta) \end{pmatrix}$$

written in the well-known 2x2 Pauli basis [49]. We note that the rotation operation acts on polarization in the case of photons, or on spin in the case of atoms or ions. In the Pauli basis the y-rotation operation is expressed as:

$$R_y(\theta) = \begin{pmatrix} \cos(\theta) & -\sin(\theta) \\ \sin(\theta) & \cos(\theta) \end{pmatrix}.$$

This unitary operation is followed by a spin- or polarization-dependent translation T, which can be mathematically expressed by:

$$T = \sum_x |x+1\rangle\langle x| \otimes |H\rangle\langle H| + |x-1\rangle\langle x| \otimes |V\rangle\langle V|,$$

with $H = (1,0)^T$ and $V = (0,1)^T$.

The quantum evolution operator for a discrete time-step is generated by a Hamiltonian $H(\theta)$, such that $U(\theta) = e^{-iH(\theta)}$ $(\hbar = 1)$, where:

$$H(\theta) = \int_{-\pi}^{\pi} dk \left[E_\theta(k) \vec{n}(k).\vec{\sigma} \right] \otimes |k\rangle\langle k|$$

and $\vec{\sigma}$ are the Pauli matrices, which readily reveals the spin-orbit coupling workings in the system. The discrete-time quantum walk described by the unitary operator $U(\theta)$ has readily been experimentally implemented in a number of devices such as photonic, cold-atom and trapped-ion devices [47, 48, 50–52]. It has been shown to display chiral symmetry and exhibit a Dirac-like dispersion relation, expressed as $\cos(E_\theta(k)) = \cos(k)\cos(\theta)$. In general, the spectrum of the system will depend on the selected branch cut. Here, we select the branch cut to be at the quasi-energy gap [53, 54].

3. Photonic DTQWs

3.1 Multiplexed DTQWs in the spatial domain

The original strategy for implementation of photonic DTQW via spatial-mode multiplexing was first introduced by Broome *et al.* [47]. The dimension of the Hilbert space for the spatial DTQW is determined by $2n + 1$ multiplexed longitudinal spatial modes of single photons coupled to a coin operation encoded in the two-dimensional spin or polarization subspace $\{|H\rangle, |V\rangle\}$. The discrete spatial modes of single photons $\{|j\rangle\}$ are labeled as $j = \pm(n - 2k)$ with $k = 0, 1, \ldots, \lfloor n/2 \rfloor$, where n denotes the walker's discrete-time step. Single-photons created via SPDC (Spontaneous Parametric Down-Conversion) in a non-linear PPKTP crystals are injected into a free-space reference spatial mode $|j\rangle = |0\rangle$. This reference mode is sequentially spatially multiplexed by a concatenation of calcite polarizing beam-displacers (CBD). Arbitrary coin states are prepared by a polarizing beam-splitter in combination with a half- (HWP) and quarter wave-plates (QWP). In due course, a combination of a HWP and a CBD implements a single discrete-step evolution. By concatenating n of such unitary arrangements one can implement n steps of a DTQW (see **Figure 1(a)** for reference). Coincident detection of photons at Avalanche Photo Detectors (APDs) (4.4 ns time window) herald a successful run of the walk. The typical number steps implemented with spatial-multiplexed schemes is of order $n \approx 10$ [47].

3.2 Multiplexed DTQW in the temporal domain

The strategy for implementation of photonic DTQW via temporal-mode multiplexing was first introduced in Ref. [48]. The dimension of the Hilbert space for the DTQW is determined by a unique spatial mode $|j\rangle = |0\rangle$ and 2^n multiplexed temporal modes $|k\rangle$ (for $k = 1, 2, \ldots, 2^n$), with n the discrete time-step number. The spatial mode is coupled to a coin operator in a two dimensional polarization subspace $(|H\rangle, |V\rangle)$ (see **Figure 1(b)**). Analogue single-photon states (on average) are generated via an attenuated pulsed diode laser. The initial states of the photons injected in the DTQW are controlled by means of half-wave plates (HWPs) and quarter-wave plates (QWPs), in order to produce eigenstates of the chirality operator $|\psi_0^{\pm}\rangle = |0\rangle \otimes (|H\rangle \pm i|V\rangle)/\sqrt{2}$. Inside the loop, the unitary rotation $(R_n(\theta))$ is implemented by a HWP with its optical axis oriented in the direction $\theta/2$. The

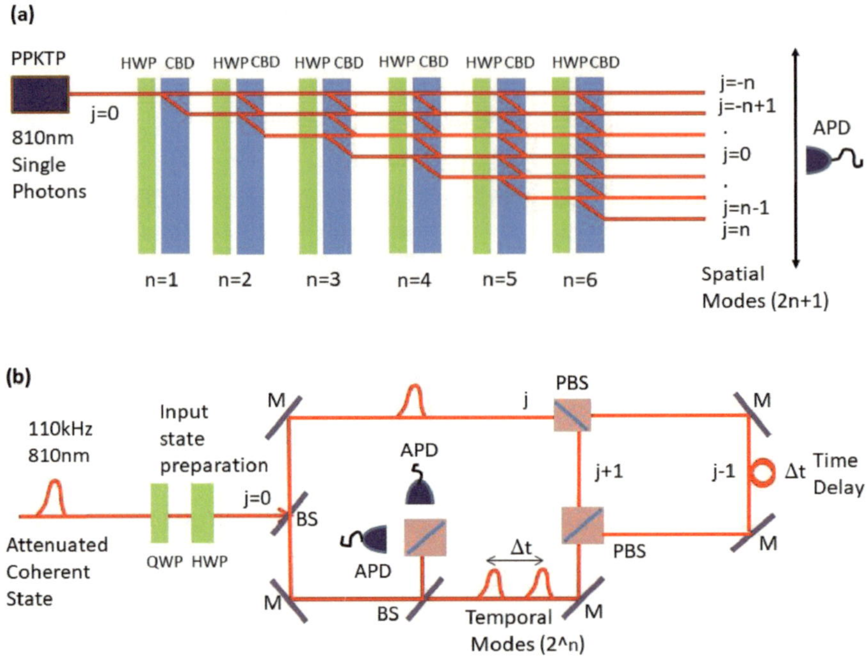

Figure 1.
Depicted architecture for experimental realization of DTQW (a) via spatial-mode multiplexing, (b) via temporal-mode multiplexing (see text for details).

polarization-dependent translation operator T is realized in the temporal domain via a polarizing beam splitter (PBS) in addition to a calibrated temporal delay-line using polarization preserving optical fibers, in which horizontally polarized light follows a longer path. The resulting calibrated temporal delay between both polarization components corresponds to a single step in the DTQW ($x \pm 1$). Polarization controllers (PCs) are introduced to compensate for arbitrary polarization rotations in the fibers. After implementing the polarization-dependent temporal difference, the so-called "time-bins" are recombined in a single spatial mode by means of a second PBS and they are directed into the fiber loops. Detection is accomplished by coupling a portion of the photons out of the loop, via a beam sampler (BS) with a probability of 5 % per step. Compensation HWPs (CHWPs) are introduced to correct for unintended dichroism introduced by the beam sampler (BS). Single-photon detectors (SPD) and avalanche photo-diodes (APDs) are employed to detect the photon arrival-time and to determine its polarization component. The probability that a photon undergoes a full round-trip is given by the overall coupling efficiency ($> 70\%$) and the total losses in the system resulting in $\eta = 0.50$. The average number of photons per input pulse is determined by neutral density (ND) filters, and is typically below $\langle n \rangle < 0.003$ to reduce contribution from multi-photon events. Such a scheme enables for implementation of a large number of discrete-time steps (typically $n \approx 20$) in a compact scheme, thus reducing the footprint characterizing spatially multiplexed architectures.

3.3 DTQW using spatial light modulators (SLM) and transverse spatial modes

We will identify the lattice points of a DTQW in a 1D geometry by the transverse spatial modes of a single photon. More specific, for photonic propagation in z-direction, the lattice sites in 1D will correspond to the position x (or y) of the transverse propagation plane. The Hilbert space of the quantum walker will be given by the discrete basis $\{|j\rangle : j \in \mathbb{Z}\}$, where $|0\rangle$ corresponds to the spatial mode aligned with the optical axis and $\{|j > 0\rangle\}$ ($\{|j < 0\rangle\}$) correspond to the upper (lower) modes, as depicted in **Figure 2(a)**. The use of transverse modes of photons for DTQW has been demonstrated for a single step by Francisco *et al.* in an intricate setup [55], encoding th subspace of the quantum coin in the upper and lower regions of the x-axis. Here, we propose a physically more intuitive approach: the quantum coin is encoded in the 2-Dimensional polarization degrees of the photon that is in the horizontal/vertical basis, i.e., $\{|H\rangle, |V\rangle\}$. In this manner, the polarization dependent translation operator T can be expressed as:

$$T = \sum_j |j+1\rangle\langle j| \otimes |H\rangle\langle H| + |j-1\rangle\langle j| \otimes |V\rangle\langle V|. \qquad (1)$$

For an unbiased coin operator, we have:

$$R = \frac{1}{\sqrt{2}}\begin{pmatrix} 1 & 1 \\ 1 & -1 \end{pmatrix}. \qquad (2)$$

Considering the following initial state for the quantum walker $|\psi_0\rangle = |0\rangle|H\rangle$, the temporal evolution of the initial quantum state after n steps will be given by:

$$|\psi_n\rangle = (TR)^n|\psi_0\rangle = \frac{1}{\sqrt{n+1}}\sum_{j=0}^{n} e^{i\phi_{n-2j}}|n-2j\rangle|\theta_{n-2j}\rangle, \qquad (3)$$

with $\phi_{n-2j} = 0$ or π, and $|\theta_{n-2j}\rangle = \cos(\theta_{n-2j})|H\rangle + \sin(\theta_{n-2j})|V\rangle$ the polarization state of the coin in the $(n-2j)$-th spatial mode. As an example, for $n = 4$:

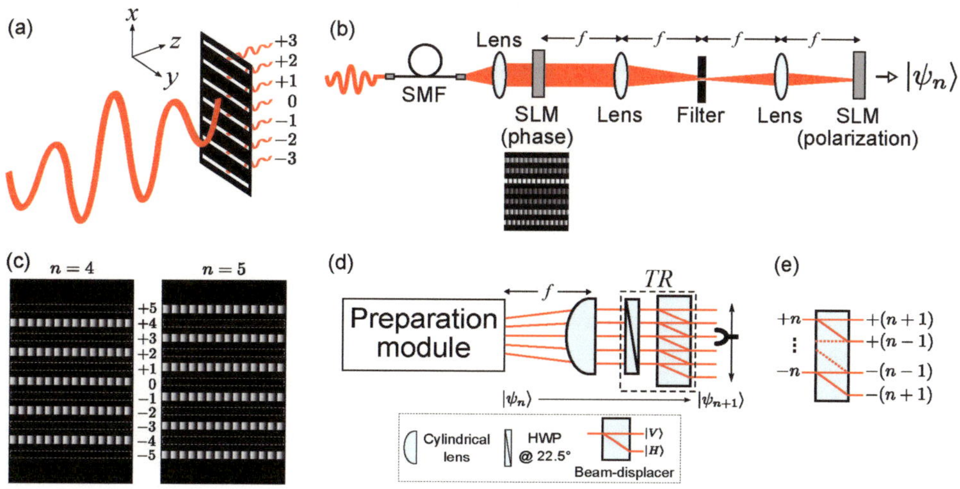

Figure 2.
(a) Discretization of a single-photon spatial amplitude profile in transverse modes along the x-direction. (b) Sketch of the proposed optical setup for preparing the n-th step walker-coin state (3) encoded in the transverse modes and polarization of a single-photon field: SMF, single mode fiber for spatial filtering; SLM, programmable spatial light modulator (see text for details). (c) Phase masks addressed at the phase-only SLM for preparing the state given by Eq. (6) with n = 4 and 5. The dashed rectangles indicate empty transverse modes. (d) Optical module that implements one step ($|\psi_n\rangle \to |\psi_{n+1}\rangle$) of the 1D DTQW proposed here (see text for details). (e) Numbering convention of the spatial modes exiting the beam-displacer [47].

$$|\psi_4\rangle \propto |+4\rangle|H\rangle + |+2\rangle\left(\frac{3|H\rangle + |V\rangle}{\sqrt{10}}\right) + |0\rangle\left(\frac{|V\rangle - |H\rangle}{\sqrt{2}}\right) - |-2\rangle\left(\frac{|V\rangle - |H\rangle}{\sqrt{2}}\right) - |-4\rangle|V\rangle. \tag{4}$$

Thus, the corresponding probability distribution characterizing the quantum walker, after n steps, will be given by:

$$P_n(j) = |(\langle H|\langle j|)|\psi_n\rangle|^2 + |(\langle V|\langle j|)|\psi_n\rangle|^2. \tag{5}$$

In order to analyze DTQW in 1D, within the framework described above, we present a realistic optical setup which can be divided into two modules. The first module is destined to prepare the initial state of Eq. (3) for an arbitrary value of n, only limited, in principle, by the resolution of the SLM used to manipulate the transverse spatial modes of photons. The second module, is destined to implement a single step in the protocol, namely, the unitary operation $U = TR$, with T and R given by Eqs. (1) and (2), respectively. With such preparation module, the probability distributions given by Eq. (5) can be measured directly. In addition, by concatenating it with the one-step propagation module, it will be possible to implement an arbitrary step in the quantum walk from n to $n+1$. Therefore, in principle it is possible to simulate 1D DTQW for arbitrary steps n, (with n a large number) surpassing the number of steps that can be implemented with time- or spatial-multiplexing approaches. In what follows, we describe the proposed preparation and propagation modules, in addition to the measurement module required to estimates the probability $P_n(j)$.

3.3.1 DTQW preparation optical module

In **Figure 2(b)** a sketch of the optical module proposed in order to prepare the input state of the quantum walker, corresponding to the n-th step of a DTQW (Eq. (3)) employing polarization degrees of freedom and transverse spatial modes of single photons. The preparation module is divided into two submodules: the first submodule, is employed to prepare the spatial degrees of freedom of the input state, and the second submodule, is employed to spatial modes with the polarization degree of freedom. A key element for the appropriate implementation of such preparation module are state-of-the-art programmable spatial light modulators (SLMs). Such SLM devices, typically based on liquid crystal display (LCD) technologies, consist of a two-dimensional array of pixels, which when properly programmed, can control the phase, amplitude and polarization of the incident light field [56]. Recently, they have been deployed in a vast number of quantum information and communication protocols [57–60].

Let us consider $\int d\vec{r}\,\psi\left(\vec{r}\right)|1\vec{r}\rangle \otimes |H\rangle$ as the quantum state of a monochromatic single-photon multi-mode field horizontally polarized in the paraxial approximation, here $\vec{r} = (x, y)$ is the position coordinate in the transverse plane, and $\psi\left(\vec{r}\right)$ is the normalized transverse probability amplitude. Such single-photon states can be generated, for example, from a spontaneous parametric down-conversion (SPDC) single photon source. The transverse amplitude $\psi\left(\vec{r}\right)$ can be manipulated using the technique developed by Prosser *et al.* [61]. Within this approach, it is possible to prepare arbitrary states of the form $\sum_j \beta_j |j\rangle$ with $\sum_j \left|\beta_j\right|^2 = 1$, where $\{|j\rangle\}$ represent the orthogonal transverse spatial modes, in the x-direction. In brief, this technique utilizes an SLM which modulates the phase information of the incident profile $\psi\left(\vec{r}\right)$ while leaving unaffected its amplitude or polarization. For simplicity, such phase information is assumed to be uniform across the entire surface of modulation. Next, a phase mask based an array of d rectangular regions, each region corresponding to a blazed diffraction grating, is displayed on the liquid crystal screen (an example of phase mask for $d = 7$ is depicted in the inset of **Figure 2(b)**). The single photon phase profile is modulated by this mask and, in the far field, light beam it is diffracted into different orders $(0, \pm 1, \dots)$ as it reaches regions with blazed gratings; otherwise, the beam propagates straight to the zeroth order. By choosing the first $+1$ order to prepare the states, the modulus of its complex coefficients will be evaluated according to the phase-modulation depth of each grating, which determines the intensity diffracted to the selected order. In addition, the phase of the coefficients will be a constant value added to the gratings. Finally, the $+1$ diffraction order is filtered by a slit diaphragm, such that the emerging photon is in a coherent superposition of d transverse "slit" modes $\{|j\rangle\}$. More specific, the states can be prepared as:

$$|\chi_n\rangle = \left(\frac{1}{\sqrt{n+1}}\sum_{j=0}^{n} e^{i\phi_{n-2j}}|n-2j\rangle\right)\otimes|H\rangle, \tag{6}$$

where $\phi_{n-2j} = 0$ or π, and n a positive integer. For a given n, one configures a phase mask for the SLM with $d = n + 1$ slit/diffraction gratings, symmetrically distributed starting from the highest modes $j = \pm n$. **Figure 2(c)** shows typical examples of masks for $n = 4$ and $n = 5$. As a technical remark, since the states we intend to prepare are uniform in phase (see Eqs. (3) and (6)), the phase-modulation depth of the gratings displayed a liquid crystal SLM will be a constant. Therefore, we can set it to be equal to 2π, ideally achieving 100% of diffraction efficiency in $+1$ order.

In order to prepare the state given by Eq. (3) starting by the input state given by Eq. (6), it is required to implement polarization rotations conditioned on the transverse-mode positions, as described by the unitary operator

$$\sum_{j=0}^{n}|n-2j\rangle\langle n-2j|\otimes\mathcal{R}(\vartheta_{n-2j}), \tag{7}$$

where

$$\mathcal{R}(\vartheta) = \begin{pmatrix} \cos\vartheta & -\sin\vartheta \\ \sin\vartheta & \cos\vartheta \end{pmatrix} \tag{8}$$

transforms $|H\rangle$ into an arbitrary state of linear polarization. By applying this rotation on $|\chi_n\rangle$ with the appropriate $\mathcal{R}(\vartheta_{n-2j})$'s, one generates the desired state $|\psi_n\rangle$ using the preparation module.

Spatially-dependent polarization rotations can be implemented by means of an SLM programmed for such task [62]. There are several different techniques for the various types of existing SLMs, which enable each pixel of the SLM device to work effectively as programmable polarization rotator [63, 64]. The details of these techniques are beyond the scope of the present work. With such programmable SLM, the transformation (7) onto the state (6) can be implemented by manipulating the transverse spatial modes of $|\chi_n\rangle$ on the liquid crystal display screen and applying proper modulation on the input polarization $|H\rangle$. As depicted in **Figure 2(b)**, this procedure is achieved using a $4f$ lens system which will image the filtered output field of the phase-only SLM onto a polarization-rotator SLM.

This concludes the description of the proposed preparation optical module for arbitrary walker-coin state in the n-th step of a 1D DTQW, encoded in polarization and transverse spatial modes, respectively, of single photons. 9 As previously states, the largest number of steps to be implemented n will be limited by the resolution of SLM. To illustrate this, consider a phase-only SLM with 2N pixels in the direction where the transverse modes are encoded (say x-direction). If each spatial mode is encoded in a row, and separated by another row of pixels, both with one-pixel width, it would be possible to define N distinguishable modes. In turn, this would enable us, in principle, to prepare the walker-coin state (3) up to $n = \lfloor N/2 \rfloor$. For a standard SLM with $2N = 1920$ [56], then $n = 4808$, which represents a much larger figure than the maximum number of steps that can be implemented with multiplexed schemes. After the preparation module, one can determine the probability distribution (5) by recording the photon count rates at each of the $2n + 1$ output transverse modes of the second SLM (see **Figure 2(b)**), appropriately normalizing to the total number of counts. This can be achieved with an array of $2n + 1$ avalanche photo-diodes (APDs) or with a single-photon detector scanning along the transverse modes. The detection apparatus has to be located right after the second SLM, in order to prevent the diffracted modes to interfere and alter the probability distribution. Alternatively, as will be described below, a transverse-to-longitudinal mode conversion can be implemented which would enable to locate the detector at greater distances from the preparation module.

3.3.2 DTQW one-step propagation module

The quantum coin operator is the quantum analogue of a walker throwing a coin, and deciding whether to proceed to the left or to right, depending on whether the coin falls heads or tail. By encoding the left and right information in the 2 dimensional photon polarization basis $|H\rangle$ and $|V\rangle$, the quantum operator corresponding to flipping a coin R, as presented in Eq. (2), can be readily implemented by using a polarization half wave plate (HWP) oriented at $\pi/4°$. Moreover, in order to implement the polarization-dependent translation operation T, as described in Eq. (1), it is straightforward to employ a birefringent element. However, this element, should also prevent the transverse modes from propagating in free space, in order to limit unwanted diffraction and interference, which would seriously hamper the characterization of the walker's translation. To maintain the discrete lattice structure of the protocol, while working with transverse modes which are properly discretized in the plane of state preparation but which are not properly discretized after the single-step due to free space propagation, one must apply a discretization procedure along all propagation planes. This can eventually be achieved by introducing a cylindrical lens with focal distance f, located at a distance f to the second SLM utilized in the preparation module, this is schemed in **Figure 2(d)**. In this way, the transverse modes at the output preparation plane are transformed into longitudinal modes, along all the remaining propagation planes. Once this transverse to longitudinal model conversion is enforced, one can simply use a polarizing calcite beam-displacer or a polarizing beam splitter in order to implement the polarization translation T. As illustrated in the inset of **Figure 2(d)**, such optical element may be oriented transmit vertically polarized light (V) and introduce a lateral beam displacement into the neighboring mode on horizontally polarized light (H). In summary, for an input state $|\psi_n\rangle$ given by (3), the one-step propagation module TR consists of a HWP oriented at $\pi/4°$ and a calcite beam-displacer in order to implement R and T, respectively, in addition to a cylindrical lens which enables transverse-to-longitudinal mode conversion. After this single-step propagation module, it possible to detect the probability distribution $P_{n+1}(j)$ (Eq. (5)), as described above. The entire procedure is depicted in **Figure 2(d)**. Furthermore, **Figure 2(e)**, describes our convention for labeling spatial modes after propagation through the calcite beam-displacer.

4. DTQW: applications in topology and geometry

Geometric phases acquired during quantum evolution of a particle can have different origins. The Berry phase [65] is a type of geometric phase that can be assigned to quantum particles which return their initial state adiabatically, while recording the path information on a geometric phase (Φ), defined as [65, 66]:

$$e^{i\Phi} = \langle \psi_{\text{ini}} | \psi_{\text{final}} \rangle. \tag{9}$$

A number of physical consequences can be attached to geometric phases, such as the modification of material properties in solids, for example the conductivity in Graphene [67, 68], the emergence of surface edge-states in topological insulators, whose surface electrons experience a geometric phase [69], the modification of molecular chemical reactions [70], and more recently geometric phases have been predicted to have implications for quantum technology, via the elusive Majorana particle [71].

In this review, we report on the progress in the characterization of geometry and topology of DTQW architectures consisting of a unitary step U given by a sequence of two non-commuting rotations in parameter space, followed by a spin-dependent translation. The topological parameter space of the DTQW architecture we analyze does not present continuous 1D topological boundaries. Unlike the "split-step" quantum walk [16, 19], or other analogous systems recently studied in the literature, the platform we report only presents a discrete number of Dirac points, where the quasi-energy gap closes. At these discrete Dirac points, the Zak Phase difference is not defined; therefore, these discrete points represent topological boundaries of zero dimension. Here we ascribe a topological boundary at the set of points where the topological invariant is not defined, namely at the discrete points where the quasi-energy gap closes. Such gapless points can be considered topological defects in parameter space. Since the system has topological defects, we argue the system is topologically non-trivial. We demonstrate the non-trivia topological l landscape of the system by calculating different holonomic and geometric quantities, such as the Zak phase, which corresponds to the Berry phase in the Brillouin zone.

5. Topology and the geometric Zak phase

The physical concept of geometric phase, such as Berry or Zak phase, is intimately linked to the concept of holonomy of a manifold. Holonomy from a geometrical standpoint: within the framework of differential geometry, an holonomy group H_x at a given point in space x for an oriented n-dimensional manifold M endowed with a given metric g_{ij} can be assigned via the (parallel) transport of a vector field $V \in TM_x$ along all possible closed curves C, starting and ending at the same point x. The condition for parallel transport is mathematically represented by the following expression:

$$t^{\mu}\nabla_{\mu}V = 0, \tag{10}$$

here t^{μ} is the tangent versor to the curve C and ∇_{μ} the Levi–Civita connection of (M, g_{ij}), representing a unique torsion free connection which satisfies $\nabla g_{\mu\nu} = 0$. By the Levi–Civita conditions together with (10), one can derive that:

$$t^{\mu}\nabla_{\mu}\left(g_{ij}V^{i}V^{j}\right) = 0,$$

stating that the vector field norm $\|V\| = g_{ij}V^{i}V^{j}$ is conserved upon travel along the closed curve C. Nevertheless, the resulting vector V_C after travel will not necessarily coincide with V, in general it will be rotated in the form:

$$V_C = R_x(C)V,$$

where $R(C)$ is an element of $SO(n)$. Therefore, a rotation matrix $R_x(C)$ corresponding to any pair (x, C) with C an arbitrary curve in the manifold can be assigned. The set of rotations $R_x(C)$ at a fixed point x can be obtained by considering all possibles curves C forming a group, which turns out to be equal or smaller than $SO(n)$. This set is known as the holonomy group H_x at the fixed point x. For simply connected manifolds M, the holonomy groups at points x and y are isomorphic. In such cases, we refer to the holonomy H of M. Otherwise, the definition of holonomy group becomes point dependent.

The geometric concept of holonomy can be depicted for a manifold M which is embedded in space R^n. An illustrative example is the sphere S^2 with its canonical metric given by:

$$g_{S^2} = d\theta^2 + \sin^2\theta d\phi^2. \tag{11}$$

This sphere represents a surface $x_1^2 + x_2^2 + x_3^2 = 1$ embedded in a space R^3. The canonical metric g_{S^2} is the distance element within this surface. For a given curve C, the holonomy element $H(C)$ is a rotation $R(\alpha)$ where $\alpha(C)$ is the solid angle subtended by the curve at the center of the sphere. It can be instructive to verify this explicitly. Consider a unit vector r in R^3, which parameterizes the points of the sphere in the form

$$r = (\sin\theta\cos\phi, \sin\theta\sin\phi, \cos\theta).$$

By taking a vector $\nu \in TM_x$, with TM_x the tangent plane to the manifold, to be transported along a curve C in S^2. By describing the travel by a parameter t, which can be interpreted as the traveling "time" along the path, the parallel transport condition can be expressed in a compact form. The vector will always be orthogonal to r, with $V \times r = 0$; if not, it would have a component orthogonal to the surface. Nevertheless, this condition is not enough since ν could reside in the tangent plane T_xS^2 of any point x along the curve. In this case, a velocity Ω with non-zero component in r-direction could make the vector rotate. In order to avoid such rotations, it is customary to insure that Ω has no components in the r-direction, a condition which be expressed as $\Omega \times r = 0$. The resulting angular velocity Ω should be different from zero; however, since ν has a conserved component in $\dot{r}$ and $\dot{r}$ produces a rotation in R^3 as well as V. These two conditions are mathematically expressed as:

$$\dot{\nu} = \Omega \times \nu, \qquad \Omega = r \times \dot{r}. \tag{12}$$

For the applications in DTQW that we intend to consider, we can express the condition above in terms of a complex unit vector ψ, defined by:

$$\psi = \frac{1}{2}(v + iv'), \qquad v' = r \times v.$$

In order to find the solid angle $\alpha(C)$, it possible to define a local orthogonal basis, with vector elements u and v.These elements are explicitly given as:

$$u(r) = (-\sin\phi, \cos\phi, 0), \tag{13}$$

$$v(r) = (-\cos\theta\cos\phi, -\cos\theta\sin\phi, \sin\theta). \tag{14}$$

On the other hand, the phase α of ψ can be expressed as:

$$\psi = n\exp(i\alpha), \qquad n = \frac{1}{2}(u + iv).$$

Note that α depends on the choice of u and v, but the phase change due to the transport along C does not. Such phase change is expressed as:

$$\alpha(C) = \oint_C d\alpha = \Im\left(\oint n^* \cdot dn\right) = \Im \iint_{Int(C)} dn \wedge dn^*,$$

the previous step makes use of the Stokes. It can be noted that the integrand is invariant under the Gauge transformations, meaning:

$$n' = n\exp(i\mu(r)).$$

This integral can be written explicitly in terms of the coordinate system, obtaining:

$$\alpha(C) = \Im \iint_{Int(C)} d\theta d\phi\left(\partial_\theta n^* \cdot \partial_\phi n - \partial_\theta n \cdot \partial_\phi n^*\right), \tag{15}$$

$$\alpha(C) = \iint_{IntC} \sin\theta d\theta d\phi, \tag{16}$$

which is the solid angle subtended by C.

Within the framework of quantum mechanics, one can replace the mathematical complex vector $\psi(\theta, \phi)$ by a quantum state vector $|\psi(X)>$, where X s are the coordinates describing the parameter space.

A complex basis $|n(X)>$ for any X can be introduced, and the relative phase of $|\psi(X)>$ can be defined:

$$|\psi\rangle = |n(X)\rangle\exp(i\gamma).$$

This phase is of course is base dependent, but the holonomy is independent from that choice of basis. Holonomy can be defined by an adiabatic travel around a curve C, in parameter space. Upon the travel, the resulting wave function accumulates an additional phase due to the non-trivial holonomy of such space. In other words,

$$\langle\psi_{\text{ini}}|\psi_{\text{final}}\rangle = \exp(i\alpha(C)).$$

The phase $\alpha(C)$ is known as the Berry phase [65]. The condition of parallel transport becomes in this context

$$\Im\langle\psi|d\psi\rangle = 0.$$

By simple generalization of the arguments given above in the differential geometrical context, it follows that this phase is simply

$$\alpha(C) = \iint_{Int(C)} \Im < dn| \wedge |dn > . \tag{17}$$

Note that the phase is dependent on the choice of path C.

The natural language for an holonomy in this context is in terms of principal bundles. There exists a natural metric g_{ij} in the parameter space of the problem. This issue was studied in [72], where the authors considered the following tensor

$$T_{ij} = < \partial_i n|(1-|n> <n|)|\partial_j n > .$$

This tensor is Gauge invariant

$$|n(X)> \rightarrow |n(X)> \exp\left(-i\mu(r)\right).$$

One may define a "distance" between two states by

$$\Delta_{12} = 1 - |<\psi_1|\psi_2>|^2.$$

The interpretation of distance is as follows. For two states $|\psi_1>$ and $|\psi_2>$ which differ only by a global phase are defined, we have $\Delta_{12} = 0$. Taking the limit $1 \rightarrow 2$ and using the fact that the states are normalized we obtain

$$ds^2 = <dn|(1-|n> <n|)|dn> = T_{ij} dX^i dX^j = g_{ij} dX^i dX^j, \tag{18}$$

this follows from the fact that the product of a symmetric tensor by an antisymmetric one is zero. Note that, for a 2-dimensional spin system:

$$|+> = \begin{pmatrix} \cos\frac{\theta}{2} e^{i\frac{\phi}{2}} \\ \sin\frac{\theta}{2} e^{-i\frac{\phi}{2}} \end{pmatrix}, \qquad |-> = \begin{pmatrix} \sin\frac{\theta}{2} e^{i\frac{\phi}{2}} \\ -\cos\frac{\theta}{2} e^{-i\frac{\phi}{2}} \end{pmatrix}$$

gives the canonical metric on S^2 (18).

A subtle remark is in order, in relation to the coloquial use of the words geometry, holonomy, and topology. The holonomy of a manifold M in a geometry context is geometrical, in the sense that the notion of parallel transport described above is related to the Levi–Civita connection ∇_i, which is constructed for a particular metric tensor g_{ij} defined on the manifold M. Nevertheless, such holonomy in general is not a topological invariant for M. More specific, two different complete metrics g_{ij} and g'_{ij} may exist, defined on the same manifold M, but possessing different holonomy groups [2]. In the context of quantum physics, the Berry phase or Zak phase may nevertheless describe topological phenomena. In fact, the description of the Berry phase above has a formal analogy with the concept of holonomy.

We can define such a geometric phase as the holonomy for an abstract connection in a principal bundle $P(U(1), X)$, where X the parameter space X. The curvature of this connection is defined in (17), also known as the Berry curvature which is Gauge invariant with flux given by Berry phase $\alpha(C)$. For closed manifolds, such fluxes describe a Chern class of the bundle. These classes take integer values and are invariant under Gauge transformations. Such classes describe different bundles in parameter space X and are topological invariant, basically meaning that they do not depend on the choice of the metric in the underlying manifold X [65].

In the following section, we present applications of these mathematical concepts within the context of quantum mechanical problems, in particular of DTQWs.

5.1 Applications via spatial multiplexing: split-step DTQW

In this Section we analyze in detail two cases of topologically non-trivial Zak phase landscape, where thee Zak phase is the equivalent to the Berry phase across the Brillouin zone. The first, the so-called split-step DTQW is implemented by applying two consecutive conditional translations T and rotations R characterized by rotation parameters $\theta_{1,2}$, such that the unitary step becomes $U(\theta_1, \theta_2) = TR(\theta_1)TR(\theta_2)$, as described in detail in [19]. The so-called "split-step" DTQW has been demonstrated to exhibit non-trivial topology characterized by distinct topological sectors, which are in turn delimited by continuous linear 1D topological boundaries. Such topological sectors are typically characterized by topological invariants, for DTQWs this is typically the winding number W, which can take binary integer values $W = 0, 1$.

The dispersion relation, which expresses the quasi-energy E as a function of the quasi-momentum k and the DTQW parameters $\theta_{1,2}$ for the split-step DTQW, results in [19]:

$$\cos\left(E_{\theta,\phi}(k)\right) = \cos(k)\cos(\theta_1)\cos(\theta_2) - \sin(\theta_1)\sin(\theta_2).$$

In order to decompose the DTQW Hamiltonian of the system in terms of Pauli matrices $H_{\mathrm{QW}} = E(k)\vec{n}\cdot\vec{\sigma}$ becomes [16], we require to know (x,y,z) components of the 3D-norm [19]:

$$\begin{aligned} n^x_{\theta_1,\theta_2}(k) &= \frac{\sin(k)\sin(\theta_1)\cos(\theta_2)}{\sin(E_{\theta_1,\theta_2}(k))}, \\ n^y_{\theta_1,\theta_2}(k) &= \frac{\cos(k)\sin(\theta_1)\cos(\theta_2) + \sin(\theta_2)\cos(\theta_1)}{\sin(E_{\theta_1,\theta_2}(k))}, \\ n^z_{\theta_1,\theta_2}(k) &= \frac{-\sin(k)\cos(\theta_2)\cos(\theta_1)}{\sin(E_{\theta_1,\theta_2}(k))}. \end{aligned} \tag{19}$$

We now turn to our second example of topologically non-trivial DTQW.

5.2 Applications via temporal multiplexed: DTQW with non-commuting rotations

As a second non-trivial example, we introduce a DTQW consisting of two sequential non-commuting rotations R_1 and R_2, which constitute the main building block of the unitary step U in the DTQW [2]. While the first rotation R_1 is performed along the y-direction by an angle θ, the second rotation R_2 is performed along the x-direction, by an angle ϕ. In this manner, the unitary step becomes $U(\theta, \phi) = TR_x(\phi)R_y(\theta)$, where $R_x(\phi)$ is also given in the Pauli basis [49] by:

$$R_x(\phi) = \begin{pmatrix} \cos(\phi) & i\sin(\phi) \\ i\sin(\phi) & \cos(\phi) \end{pmatrix}.$$

The 3D-norm required for expressing the Hamiltonian in the Pauli basis, results in:

$$\begin{aligned} n^x_{\theta,\phi}(k) &= \frac{-\cos(k)\sin(\phi)\cos(\theta) + \sin(k)\sin(\theta)\cos(\phi)}{\sin\left(E_{\theta,\phi}(k)\right)}, \\ n^y_{\theta,\phi}(k) &= \frac{\cos(k)\sin(\theta)\cos(\phi) + \sin(k)\sin(\phi)\cos(\theta)}{\sin\left(E_{\theta,\phi}(k)\right)}, \\ n^z_{\theta,\phi}(k) &= \frac{-\sin(k)\cos(\theta)\cos(\phi) + \cos(k)\sin(\theta)\sin(\phi)}{\sin\left(E_{\theta,\phi}(k)\right)}. \end{aligned} \tag{20}$$

The dispersion relation for the DTQW with non-commuting rotations results in:

$$\cos\left(E_{\theta,\phi}(k)\right) = \cos(k)\cos(\theta)\cos(\phi) + \sin(k)\sin(\theta)\sin(\phi), \tag{21}$$

it cam be easily verified that we recover a Dirac-like dispersion relation for $\phi = 0$, as expected.

As readily mentioned, the described system exhibits a non-trivial phase diagram consisting of a large number of discrete gapless points for different quasi-momenta. Such singular points can be regarded as topological defects in parameter space. Each gapless points represent topological boundaries of dimension zero, where topological invariant, such as the winding number W, are not defined. As anticipated, in contrast to the "split-step" DTQW described in previous sections, this system does not contain continuous topological boundaries. We calculated analytically the gapless Dirac points and zero-dimension topological boundaries for the system by using basic trigonometric considerations. It can be readily demonstrated that there are 13 discrete points for different values of quasi-momentum k where the gap closes. This is depicted in **Figure 3**. Different symbols correspond to different values

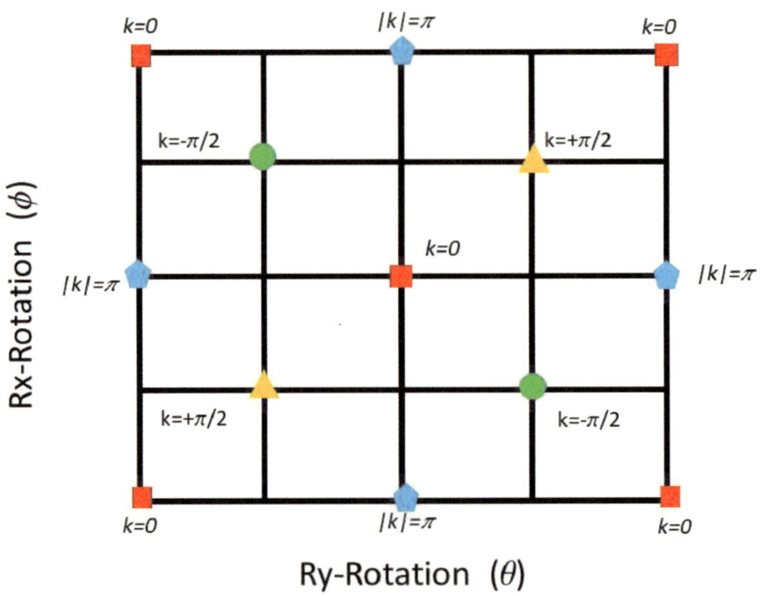

Figure 3.
Phase diagram for DTQW with non-commuting rotations. The symbols indicate gapless Dirac points where quasi-energy gap closes for different values of quasi-momentum: Squares ($k = 0$), pentagons ($|k| = \pi$), romboids ($k = +\pi/2$), and circles ($k = -\pi/2$). The discrete Dirac points represent topological boundaries of dimension zero, and endow the system with a non-trivial topology [2].

of quasi-momenta. Namely, pentagons correspond to Dirac points for $|k| = \pi$, romboids correspond to Dirac points for $k = +\pi/2$, squares correspond to Dirac points for $k = 0$, and circles correspond to Dirac points for $k = -\pi/2$. Such holonomic structure in itself is topologically non-trivial, and was studied in [2] for the first time.

6. Geometric phase calculation

We now provide expressions for the geometric phase, the so-called Zak phase acquired due to quantum evolution across the Brillouin Zone, in the two aforementioned scenarios. These two scenarios are characterized by a generic Hamiltonian of the form:

$$H \sim n_x\sigma_x + n_y\sigma_y + n_z\sigma_z. \tag{22}$$

The specific Hamiltonians for each scenario differ by a constant factor, and by the specific expressions of the normal vector n_i (with $i = x, y, z$). Since the eigenvectors of the Hamiltonian are the only quantities of interest for the present problem, overall constants can be safely ignored.

In general the Hamiltonian in the Pauli basis is given by the following matrix:

$$H = \begin{pmatrix} n_z & n_x - in_y \\ n_x + in_y & -n_z \end{pmatrix} \tag{23}$$

and is characterized by the eigenvalues, which represent the eigenenergies of the system:

$$\lambda = \pm\sqrt{n_x^2 + n_y^2 + n_z^2}. \tag{24}$$

By diagonalizing this generic Hamiltonian, we find thee normalized eigenvectors for the generic Hamiltonian are given by:

$$|V_\pm\rangle = \begin{pmatrix} \dfrac{n_x + in_y}{\sqrt{2n_x^2 + 2n_y^2 + 2n_z^2 \mp 2n_z\sqrt{n_x^2 + n_y^2 + n_z^2}}} \\ \dfrac{n_z \mp \sqrt{n_x^2 + n_y^2 + n_z^2}}{\sqrt{2n_x^2 + 2n_y^2 + 2n_z^2 \mp 2n_z\sqrt{n_x^2 + n_y^2 + n_z^2}}} \end{pmatrix}. \tag{25}$$

It is to be noted that the scaling factor $n_i \rightarrow \lambda n_i$ does not affect the result. As mentioned, this results from the fact that two Hamiltonians differing by a constant have the same eigenvectors.

The geometric Zak phase ($\Phi_{Zak} = Z$) for the positive and negative bands ($\pm$), is expressed as:

$$Z_\pm = i\int_{-\pi/2}^{\pi/2} dk \langle V_\pm|\partial_k V_\pm\rangle. \tag{26}$$

We will now apply these concepts to the specific examples reviewed in the previous sections.

6.1 Split-step DTQW

We will calculate the Zak phase for two types of DTQW, the first one is the so-called split-step DTQW [19, 73]. It consists of a DTQW with unitary step U given by the following expression $U(\theta_1, \theta_2) = TR(\theta_1)TR(\theta_2)$. Such unitary step can be readily implemented via spatial multiplexing, as described in [19, 73]. For the unitary step characterizing the split-step DTQW, the components of the normal vector n_i for decomposing the Hamiltonian in terms of Pauli operators can be written in the following manner:

$$n^x_{\theta_1,\theta_2}(k) = \frac{\sin(k)\sin(\theta_1)\cos(\theta_2)}{\sin(E_{\theta_1,\theta_2}(k))},$$
$$n^y_{\theta_1,\theta_2}(k) = \frac{\cos(k)\sin(\theta_1)\cos(\theta_2) + \sin(\theta_2)\cos(\theta_1)}{\sin(E_{\theta_1,\theta_2}(k))}, \tag{27}$$
$$n^z_{\theta_1,\theta_2}(k) = \frac{-\sin(k)\cos(\theta_2)\cos(\theta_1)}{\sin(E_{\theta_1,\theta_2}(k))}.$$

In particular, we consider the case in which the normal vector $n = (n_x, n_y, n_z)$ is fully transversal, meaning that $n_z = 0$. By setting the angle parameters such that $n_z = 0$, it can be easily demonstrated that the normalized Hamiltonian eigenvectors are of the form:

$$|V_\pm\rangle = \frac{1}{\sqrt{2}}\begin{pmatrix} e^{-i\phi(k)} \\ \mp 1 \end{pmatrix}, \qquad \tan\phi(k) = \frac{n_y}{n_x}. \tag{28}$$

There are two possible angle choices that lead to $n_z = 0$, these are $\theta_1 = 0$ or $\theta_2 = 0$. For either of these angle choices, the Zak phase for the positive and negative band take equivalent values, of the form [2]:

$$Z = Z_\pm = i\int_{-\pi/2}^{\pi/2} dk\langle V_\pm|\partial_k V_\pm\rangle, \tag{29}$$

$$Z = i\int_{-\pi/2}^{\pi/2} dk\langle V_\pm|\partial_k V_\pm\rangle = \phi(-\pi/2) - \phi(\pi/2), \tag{30}$$

from where it follows that

$$Z = \frac{\tan(\theta_2)}{\tan(\theta_1)}. \tag{31}$$

A numerical simulation of the Zak phase for the split-step DTQW is depicted in **Figure 4a**.

6.2 DTQW with non-commuting rotations

The particular DTQW with non-commuting rotations presented in previous Sections can be readily implemented via temporal multiplexing approaches. To this end, we recall that the unitary step results in $U(\theta, \phi) = TR_x(\phi)R_y(\theta)$. The Cartesian components of the 3D-norm n_i $(i = x, y, z)$ are as follows:

$$n_x = -\cos(k)a + \sin(k)b, \tag{32}$$

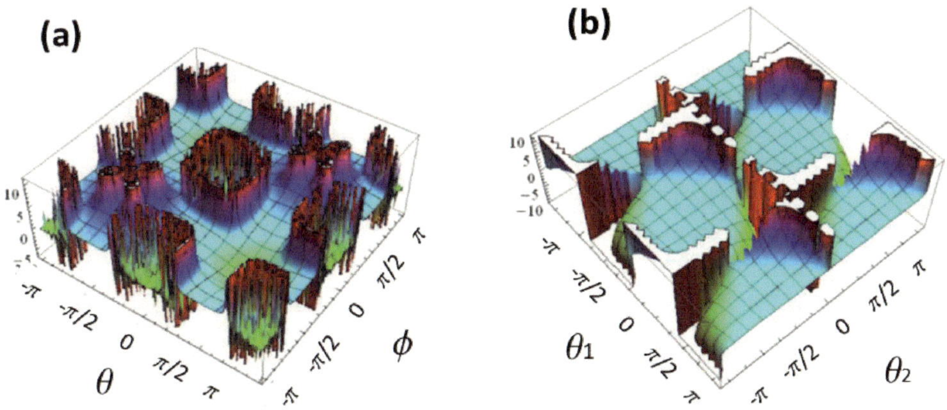

Figure 4.
(a) Non-trivial geometric Zak phase landscape for DTQW with non-commuting rotations obtained by numeric integration, (b) Geometric Zak phase landscape for "split-step" DTQW obtained analytically [2].

$$n_y = \cos(k)b + \sin(k)a, \tag{33}$$

$$n_z = \cos(k)c - \sin(k)d, \tag{34}$$

where

$$a = \sin(\phi)\cos(\theta), \tag{35}$$

$$b = \cos(\phi)\sin(\theta), \tag{36}$$

$$c = \sin(\phi)\sin(\theta), \tag{37}$$

$$d = \cos(\phi)\cos(\theta), \tag{38}$$

angular functions as defined above. N_1 is can be expressed as:

$$N_1 = n_x + in_y = -\exp(-ik)(a - ib). \tag{39}$$

In this scenario, calculation of Zak phase in terms of the Hamiltonian eigenvectors ($|V_\pm\rangle$) for each band (positive and negative) can be accomplished, resulting in:

$$Z = Z_\pm = i\int_{-\pi/2}^{\pi/2} dk\langle V_\pm|\partial_k V_\pm\rangle,$$

Making use of expression (41), the geometric Zak phase results in:

$$Z_\pm = \int \frac{(a^2 + b^2)dk}{D_\pm^2}, \tag{40}$$

$$\begin{aligned} D_\pm &= \sqrt{2n_x^2 + 2n_y^2 + 2n_z^2 \mp 2n_z\sqrt{n_x^2 + n_y^2 + n_z^2}} \\ &= \Big(a^2 + b^2 + c^2\cos^2(k) + d^2\sin^2(k) - \sin(2k)cd \\ &\quad \mp(\cos(k)c - \sin(k)d) \\ &\quad \times\sqrt{a^2 + b^2 + c^2\cos^2(k) + d^2\sin^2(k) - \sin(2k)cd}\Big)^{\frac{1}{2}}. \end{aligned} \tag{41}$$

Note that, for the case of DTQW with non-commuting rotations, the consequences of setting the norm to be fully transverse (i.e., $n_z = 0$) are quite different than in the case of the split-step DTQW. More specific, $n_z = 0$ returns a trivially constant Zak phase $Z = \pi$, since the k-dependence vanishes. For this system, there is no analytic expression for the Zak phase, and the Zak phase landscape can only be obtained by numerical integration. Note that, at the Dirac points indicated in **Figure 5**, the Zak phase is ill defined. A numerical simulation of the Zak phase Φ_{Zak} by numeric integration in Wolfram Mathematica is depicted in **Figure 4**, corresponding to parameter values of the form $\theta_{1,2} = [-\pi, \pi]$ and $\phi = [-\pi, \pi]$—(a - left) Zak phase for split-step DTQW, given by the analytic expression $Z = \frac{\tan(\theta_2)}{\tan(\theta_1)}$; (b-right) Zak phase for DTQW with non-commuting rotation, obtained by numerical integration of expression Eq. (41).

A brief discussion is in order, it is well known that the Zak phase is Gauge dependent —that is, it depends on the particular choice of origin of the unit cell [74]. Therefore, in general it is not uniquely defined and cannot be considered a topological invariant. Nevertheless, a related topological invariant quantity can be defined in terms of the Zak phase *difference* between two states $(|\psi^1\rangle, |\psi^2\rangle)$ differing by a geometric phase only. The Zak phase difference between two such states can be expressed as $\langle\psi^1|\psi^2\rangle = e^{i|\Phi^1_{Zak} - \Phi^2_{Zak}|}$. More explicit, the term geometric invariance refers to geometric properties that do not depend on the choice of origin of the Brillouin zone, and only depend on relative distances between geometric points in the Brillouin zone.

A time-multiplexed experimental scheme, which can be readily implemented to obtain the Zak phase difference between two states at a given time-step N is suggested. For a given choice of origin of the Brillouin zone, the system is characterized by a unitary evolution operator consisting of rotation parameters corresponding to either of the four adjacent Dirac points, where the gap closes. A different geometric phase will be accumulated at each adjacent Dirac point. Such phase difference can be experimentally determined by coherently recombining the

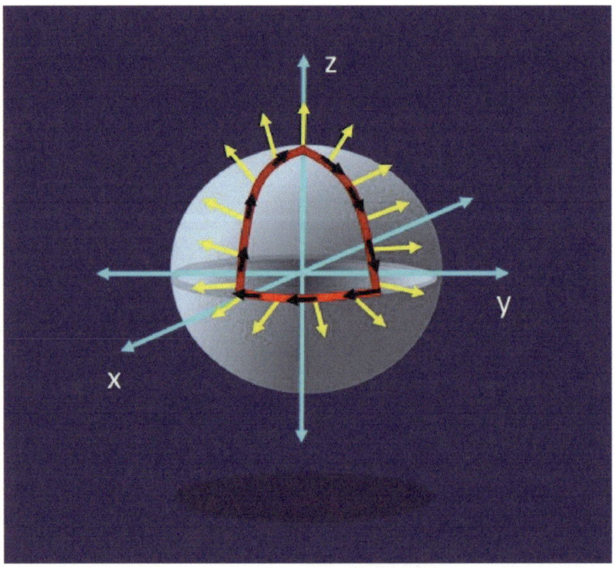

Figure 5.
Graphic depiction of the geometric phase Φ acquired along propagation in a closed trajectory. The spin or polarization (yellow arrows) remains perpendicular to the direction of propagation (black arrows). See section 4 and section 5 for further details on topology and holonomy in quantum systems.

states. More specific, in the photonic case, by interfering the states by using a Mach-Zehnder interferometer. A suitable experimental scheme for detection of the Zak phase difference in a photonic system is readily presented in [2, 75].

7. Conclusions

In this Book Chapter, we reported a review of novel approaches to photonic discrete-time quantum walk (DTQW) platforms. Namely, we discussed implementations via spatial-multiplexing or temporal-multiplexing schemes, and we introduced a novel scheme for implementations based on transverse spatial modes of photons, which are in turn controlled by spatial light modulators (SLMs). While the number of discrete time-steps (n) that can be experimentally implemented via mode multiplexed approaches is typically limited by the mode scaling of the multiplexed technique itself, i.e., $2n + 1$ for spatial mode multiplexing and 2^n for temporal-mode multiplexing, realizations using transverse modes can in principle enable experimental simulation of an arbitrary temporal step n, only limited by the resolution of the SLM itself. We present several relevant applications of DTQWs in quantum simulation. Namely, for the simulation of topological effects, ascribed to each DTQW platform. Specifically, in the context of mode-multiplexed DTQWs, we presented in detail the calculation of the Zak Phase, corresponding to the Berry phase across the Brillouin zone, for the case of the *split-step* DTQW and for the case of DTQW with non-commuting rotations, which are implemented via spatial and temporal mode-multiplexing, respectively.

Acknowledgements

The author gratefully acknowledges Leonardo Neves, Osvaldo Santillan, and Mohammad Hafezi. G.P. gratefully acknowledges financial support from PICT2015-0710 grant, and UBACYT PDE 2017 Raices programme.

Conflicts of interest

The author declares no conflict of interest.

Author details

Graciana Puentes[1,2]

1 Departamento de Física, Facultad de Ciencias Exactas y Naturales, Universidad de Buenos Aires, Ciudad Universitaria, Buenos Aires, Argentina

2 Instituto de Física de Buenos Aires (IFIBA), Universidad de Buenos Aires-CONICET, Ciudad Universitaria, Buenos Aires, Argentina

*Address all correspondence to: gpuentes@df.uba.ar

References

[1] This review is based on two articles published by the Author [2, 3] under Creative Commons Attribution License which permits unlimited reproduction, upon proper citation.

[2] G. Puentes, Topology and holonomy in discrete-time quantum walks, *Crystals* 7, 122 (2017); selected for Cover Story of a Special Issue on Quantum Topology (Open Access Article, distributed under the Creative Commons Attribution License which permits unrestricted use, distribution and reproduction provided proper citation).

[3] L. Neves and G. Puentes, Photonic Discrete-time Quantum Walks and Applications, *Entropy* **20**, 731 (2018) (Open Access Article, distributed under the Creative Commons Attribution License which permits unrestricted use, distribution and reproduction provided proper citation).

[4] A.M. Childs, R. Cleve, E. Deotto, E. Farhi, S. Gutmann, and D. Spielman. Exponential algorithmic speedup by quantum walk. In Proceedings of the 35th ACM Symposium on The Theory of Computation (STOC'03) ACM, pp. 59–68, 2003.

[5] A. Ambainis. Quantum random walks, a new method for designing quantum algorithms. In SOFSEM 2008: Theory and Practice of Computer Science, Lecture Notes in Computer Science, vol. 4910, pp. 1–4, Springer Berlin/Heidelberg, 2008.

[6] N. Shenvi, J. Kempe, and R.B. Whaley. A quantum random walk search algorithm. Phys. Rev. A, 67(5): 052307, 2003.

[7] A. Ambainis. Quantum walk algorithm for element distinctness. In Proceedings of 45th Annual IEEE Symposium on Foundations of Computer Science (FOCS'04), pp. 22–31, 2004

[8] A. Ambainis. Quantum walks and their algorithmic applications. International Journal of Quantum Information, vol. 1 (4), pp. 507–518, 2003.

[9] M. Mohseni, P. Rebentrost, S. Lloyd, and A. Aspuru-Guzik. Environment-assisted quantum walks in energy transfer of photosynthetic complexes. J Chem Phys., 129(17):174106, 2008.

[10] A.M. Childs. Universal computation by quantum walk. Phys. Rev. Lett., 102: 180501, 2009.

[11] A. Tulsi. Faster quantum-walk algorithm for the two-dimensional spatial search. Phys. Rev. A, 78(1): 012310, 2008.

[12] N.B. Lovett, S. Cooper, M. Everitt, M. Trevers, and V. Kendon. Universal quantum computation using the discrete-time quantum walk. Phys. Rev. A, 81(4):042330, 2010.

[13] M. S. Underwood and D.L. Feder. Universal quantum computation by discontinuous quantum walk. Phys. Rev. A, 82(4):042304, 2010.

[14] Kempe, J. Quantum random walks: an introductory overview. *Cont. Phys.* **2003**, 44, 307, doi:10.1080/00107151031000110776.

[15] Y. Aharonov, L. Davidovich, and N. Zagury, Quantum random walks. *Phys. Rev. A* **48**, 1687 (1993).

[16] T. Kitagawa, Matthew A. Broome, Alessandro Fedrizzi, Mark S. Rudner, Erez Berg, Ivan Kassal, Alán Aspuru-Guzik, Eugene Demler, and Andrew G. White, Observation of topologically protected bound states in photonic quantum walks. *Nature Communications* **3**, 882 (2012).

[17] A. Crespi, R. Osellame, R. Ramponi, V. Giovannetti, R. Fazio, Linda Sansoni,

F. De Nicola, F. Sciarrino, and P. Mataloni, Anderson localization of entangled photons in an integrated quantum walk. *Nature Photon.* 7, 322 (2013).

[18] Maximilian Genske, Wolfgang Alt, Andreas Steffen, Albert H. Werner, Reinhard F. Werner, Dieter Meschede, and Andrea Alberti, Electric quantum walks with individual atoms. *Phys. Rev. Lett.* **110**, 190601 (2013).

[19] T. Kitagawa, M. S. Rudner, E. Berg, and E. Demler, Exploring topological phases with quantum walks. *Phys. Rev. A* **82**, 033429 (2010).

[20] H. Obuse and N. Kawakami, Topological phases and delocalization of quantum walks in random environments. *Phys. Rev. B* **84**, 195139 (2011).

[21] C. Cedzich, F. Grunbaum, C. Stahl, L. Velázquez, A. Werner, and R. Werner, Bulk-edge correspondence of one-dimensional quantum walks. *J. Phys. A: Math. Theor.* **49**, 21LT01 (2016).

[22] Antoni Wojcik, Tomasz Luczak, Paweł, Kurzynski, Andrzej Grudka, Tomasz Gdala, and Malgorzata Bednarska-Bzdega, Trapping a particle of a quantum walk on the line.*Phys. Rev. A* **85**, 012329 (2012).

[23] Y. Shikano, K. Chisaki, E. Segawa, and N. Konno, Emergence of randomness and arrow of time in quantum walks. *Phys. Rev. A* **81**, 062129 (2010).

[24] J. K. Asbóth, Symmetries, topological phases, and bound states in the one-dimensional quantum walk. *Phys. Rev. B* **86**, 195414 (2012).

[25] H. Obuse,J. K. Asboth, Y. Nishimura, N. Kawakami, Unveiling hidden topological phases of a one-dimensional Hadamard quantum walk. *Phys. Rev. B* **2015**, *92*, 045424.

[26] S. Moulieras, M. Lewenstein, and G. Puentes, Entanglement engineering and topological protection in discrete-time quantum walks. *J. Phys. B* **46**, 104005 (2013).

[27] C. Beenakker and L. Kouwenhoven, A road to reality with topological superconductors. *Nature Phys.* **12**, 618 (2016).

[28] Meng Xiao, Guancong Ma, Zhiyu Yang, Ping Sheng, Z. Q. Zhang, and C. T. Chan, Geometric phase and band inversion in periodic acoustic systems. *Nature Phys.* **11**, 240 (2015).

[29] S. D. Huber, Topological mechanics. *Nature Phys.* **12**, 621 (2016).

[30] V. Peano, C. Brendel, M. Schmidt, F. Marquardt, Topological phases of sound and light. *Phys. Rev. X* **5**, 031011 (2015).

[31] L. Lu, J. Joannopoulos, M. Solijacic, Topological states in photonic systems. *Nature Phys.* **12**, 626 (2016).

[32] S. Bose, Quantum communication through an unmodulated spin chain. *Phys. Rev. Lett.* **91**, 207901 (2003).

[33] M. Christandl, N. Datta, A. Ekert, and A. J. Landahl, Perfect state transfer in quantum spin networks. *Phys. Rev. Lett.* **92**, 187902 (2004).

[34] M. B. Plenio, and S. F. Huelga, Dephasing-assisted transport: quantum networks and biomolecules. *New. J. Phys.* **10**, 113019 (2008).

[35] J. Spring *et al.*, Boson sampling on a photonic chip. *Science* **339**, 798–801 (2013).

[36] Matthew A. Broome, Alessandro Fedrizzi, Saleh Rahimi-Keshari, Justin Dove, Scott Aaronson, Timothy C. Ralph, Andrew G. White, Photonic boson sampling in a tunable circuit. *Science* **339**, 6121 (2013).

[37] Max Tillmann, Borivoje Dakic, René Heilmann, Stefan Nolte, Alexander Szameit and Philip Walther, Experimental boson sampling. *Nature Photon.* 7, 540 (2013).

[38] A. Crespi *et al.*,Integrated multimode interferometers with arbitrary designs for photonic boson sampling. *Nature Photon.* 7, 545 (2013).

[39] N. Spagnolo *et al.*, Efficient experimental validation of photonic boson sampling against the uniform distribution. *Nature Photonics* **8**, 615 (2014).

[40] J. Carolan *et al.*, On the experimental verification of quantum complexity in linear optics. *Nature Photonics* **8**, 621 (2014).

[41] A. M. Childs, Universal computation by quantum walk. *Phys. Rev. Lett.* **102**, 180501 (2009).

[42] A Aiello, G Puentes, D Voigt, JP Woerdman, Maximally entangled mixed-state generation via local operations. *Phys. Rev. A* 75 (6), 062118 (2007).

[43] G. Puentes, D. Voigt, A. Aiello, and J. P. Woerdman, Universality in depolarized light scattering. *Opt. Lett.* **30** (23), 3216–3218 (2006).

[44] A. Peruzzo *et al.*, Quantum walks of correlated photons. *Science* **329**, 1500–1503 (2010).

[45] K. Poulios *et al.*, Quantum walks of correlated photon pairs in two-dimensional waveguide arrays. *Phys. Rev. Lett.* **112**, 143604 (2014).

[46] Andreas Schreiber, Aurel Gabris, Peter P. Rohde, Kaisa Laiho, Martin Stefanak, Vaclav Potocek, Craig Hamilton, Igor Jex, Christine Silberhorn, A 2D quantum walk simulation of two-particle dynamics. *Science* **336**, pp. 55–58 (2012).

[47] Broome, M.A.; Fedrizzi, A.; Lanyon, B.P.; Kassal, I.; Aspuru-Guzik, A.; White, A.G. Discrete Single-Photon Quantum Walks with Tunable Decoherence. *Phys. Rev. Lett.* **2010**, 104, 153602, doi:10.1103/PhysRevLett.104.153602.

[48] A. Schreiber, K. N. Cassemiro, V. Potocek, A. Gabris, I. Jex, and Ch. Silberhorn, Decoherence and disorder in quantum walks: from ballistic spread to localization. *Phys. Rev. Lett.* **106**, 180403 (2011).

[49] M. Nielsen, and I. Chuang, *Quantum Computation and Quantum Information*, Cambridge University Press (2000).

[50] A. Schreiber, K. N. Cassemiro, V. Potocek, A. Gabris, P. J. Mosley, E. Andersson, I. Jex, and Ch. Silberhorn, Photons walking the line: a quantum walk with adjustable coin operations. *Phys. Rev. Lett.* **104**, 050502 (2010).

[51] F. Zahringer, G. Kirchmair, R. Gerritsma, E. Solano, R. Blatt, and C. F. Roos, Realization of a quantum walk with one and two trapped ions. *Phys. Rev. Lett.* **104**, 100503 (2010).

[52] Maximilian Genske, Wolfgang Alt, Andreas Steffen, Albert H. Werner, Reinhard F. Werner, Dieter Meschede, and Andrea Alberti, Electric Quantum Walks with Individual Atoms. *Phys. Rev. Lett.* **110**, 190601 (2013).

[53] Fruchart, M. *Phys. Rev. B* **2015**, *92*, 045424.

[54] M. S. Rudner, N. H. Lindner, E. Berg, M. Levin, Anomalous edge states and the bulk-edge correspondence for periodically driven two-dimensional systems. *Phys. Rev. X* **2013**, *3*, 031005.

[55] Francisco, D.; Iemmi, C.; Paz, J.P.; Ledesma, S. Simulating a quantum walk with classical optics. *Phys. Rev. A* **2006**, 74, 052327, doi:10.1103/PhysRevA.74.052327.

[56] Lazarev, G.; Hermerschmidt, A.; Krueger, S.; Osten, S. LCOS spatial light modulators: trends and applications. In *Optical Imaging and Metrology: Advanced Technologies*, 1st ed.; Osten, W., Reingand, N., Eds.; Wiley-VCH: Weinheim, Germany, 2012; pp. 1–29; ISBN: 978–3–527-41064-4.

[57] G. Puentes, C. La Mela, S. Ledesma, C. Iemmi, J.P. Paz, M. Saraceno, Optical simulation of quantum algorithms using programmable liquid-crystal displays, Physical Review A **69**, 042319 (2004).

[58] Marques, B.; Matoso, A.A.; Pimenta, W.M.; Gutiérrez-Esparza, A.J.; Santos, M. F.; Pádua, S. Experimental simulation of decoherence in photonics qudits. *Sci. Rep.* **2015**, 5, 16049, doi:10.1038/srep16049.

[59] Solís-Prosser, M.A.; Fernandes, M. F.; Jiménez, O.; Delgado, A.; Neves, L. Experimental Minimum-Error Quantum-State Discrimination in High Dimensions. *Phys. Rev. Lett.* **2017**, 118, 100501, doi:10.1103/PhysRevLett. 118.100501.

[60] Bouchard, F.; Fickler, R.; Boyd, R. W.; Karimi, E. High-dimensional quantum cloning and applications to quantum hacking. *Sci. Adv.* **2017**, 3, e1601915, doi:10.1126/sciadv.1601915.

[61] Solís-Prosser, M.A.; Arias, A.; Varga, J.J.M.; Rebón, L.; Ledesma, S.; Iemmi, C.; Neves, L. Preparing arbitrary pure states of spatial qudits with a single phase-only spatial light modulator. *Opt. Lett.* **2013**, 38, 4762, doi:10.1364/ OL.38.004762.

[62] Moreno, I.; Velásquez, P.; Fernández-Pousa, C.R.; Sánchez-López, M.M.; Mateos, F. Jones matrix method for predicting and optimizing the optical modulation properties of a liquid-crystal display. *J. Appl. Phys.* **2003**, 94, 3697, doi:10.1063/1.1601688.

[63] Davis, J.A.; McNamara, D.E.; Cottrell, D.M.; Sonehara, T. Two-dimensional polarization encoding with a phase-only liquid-crystal spatial light modulator. *Appl. Opt.* **2000**, 39, 1549, doi:10.1364/AO.39.001549.

[64] Moreno, I.; Martínez, J.L.; Davis, J. A. Two-dimensional polarization rotator using a twisted-nematic liquid-crystal display. *Appl. Opt.* **2007**, 46, 881, doi: 10.1364/AO.46.000881.

[65] M. V. Berry, Classical adiabatic angles and quantal adiabatic phase. *J. Phys. A* **18**, 15 (1985).

[66] J. Hannay, Angle variable holonomy in adiabatic excursion of an integrable Hamiltonian. *J. Phys. A* **18**, 221 (1985).

[67] Y. Zhang, Y.-W. Tan, H. L. Stormer, and P. Kim, Experimental observation of the quantum Hall effect and Berry's phase in graphene. *Nature* **438** 201 (2005).

[68] P. Delplace, D. Ullmo, G. Montambaux, Zak phase and the existence of edge states in graphene. *Phys. Rev. B* **2011**, *84*, 195452.

[69] C. L. Kane, and E. J. Mele, Z 2 topological order and the quantum spin Hall effect, *Phys. Rev. Lett.* **95**, 146802 (2005); B. Bernevig *et al.*, Quantum spin Hall effect and topological phase transition in HgTe quantum wells, *Science* **314**, 1757 (2006); M. Köning *et al.*, Quantum spin Hall insulator state in HgTe quantum wells, *Science* **318**, 766 (2007).

[70] G. Delacretaz, E. R. Grant, R.L. Whetten, L. Wöste, and J. W. Zwanziger, Fractional quantization of molecular pseudorotation in Na 3. *Phys. Rev. Lett.* **56**, 2598 (1986).

[71] S. Nadj-Perge, I. K. Drozdov1, J. Li, H. Chen, S. Jeon, J.l Seo, A. H. MacDonald, B. A. Bernevig, A. Yazdani, Observation of Majorana fermions in ferromagnetic atomic chains on a superconductor. *Science* **346**, 602 (2014).

[72] J. Provost and G. Vallee, Riemannian structure on manifolds of quantum states. Comm. Math. Phys **76**, 289 (1980).

[73] G. Puentes, Spontaneous parametric downconversion and quantum walk topology, JOSA B **33**, 461–467 (2016).

[74] M. Atala, M. Aidelsburger, J. Barreiro, D. Abanin, T. Kitagawa, E. Demler, I. Bloch, Direct measurement of the Zak phase in topological Bloch bands. *Nature Phys.* **9**, 795 (2013).

[75] J. C. Loredo, M. A. Broome, D. H. Smith, and A. G. White, Observation of entanglement-dependent two-particle holonomic phase. *Phys. Rev. Lett.* **112**, 143603 (2014).

Chapter 5

Introduction to Quantum Computing

Surya Teja Marella and Hemanth Sai Kumar Parisa

Abstract

Quantum computing is a modern way of computing that is based on the science of quantum mechanics and its unbelievable phenomena. It is a beautiful combination of physics, mathematics, computer science and information theory. It provides high computational power, less energy consumption and exponential speed over classical computers by controlling the behavior of small physical objects i.e. microscopic particles like atoms, electrons, photons, etc. Here, we present an introduction to the fundamental concepts and some ideas of quantum computing. This paper starts with the origin of traditional computing and discusses all the improvements and transformations that have been done due to their limitations until now. Then it moves on to the basic working of quantum computing and the quantum properties it follows like superposition, entanglement and interference. To understand the full potentials and challenges of a practical quantum computer that can be launched commercially, the paper covers the architecture, hardware, software, design, types and algorithms that are specifically required by the quantum computers. It uncovers the capability of quantum computers that can impact our lives in various viewpoints like cyber security, traffic optimization, medicines, artificial intelligence and many more. At last, we concluded all the importance, advantages and disadvantages of quantum computers. Small-scale quantum computers are being developed recently. This development is heading towards a great future due to their high potential capabilities and advancements in ongoing research. Before focusing on the significances of a general-purpose quantum computer and exploring the power of the new arising technology, it is better to review the origin, potentials, and limitations of the existing traditional computing. This information helps us in understanding the possible challenges in developing exotic and competitive technology. It will also give us an insight into the ongoing progress in this field.

Keywords: quantum computing, real-time systems, program processors

1. Introduction

1.1 History of computing

Evolution in one region of science and technology leads to the discovery of a new one. In less than a century, research and development of functional computing technologies have renovated science, technology, and nation massively. The first practical computer around the 20th century was not capable of doing mathematical computations, on its own. Practical devices need a solid physical implementation of theoretical concepts. Nowadays, computers are solving problems instantly and accurately provided the input is relevant, and a set of instructions given are favorable. It all started from World War II when Alan Turing created a real general-purpose computer with a storable program model and is known as the 'Universal Turing Machine'. It was redesigned by Von Neumann and is now the most important architecture for almost every computer. The computers and their physical parts kept improving with time in terms of performance and their strengths. And gradually, the industry of computers became larger than the military department which initiated it. The advancement in control and understanding of humans over nature and physical systems has given us the latest electronic devices we are utilizing today [1].

2. A new kind of computing

Today's computers are smaller, cheaper, faster, greatly efficient, and even more powerful as compared to early computers that used to be huge, costly, and more power-consuming. It becomes possible due to improvements in architecture, hardware components, and software running on them. Electronic circuits used in computers are getting smaller and smaller day by day. Transistors are small semiconductor devices that are used to amplify and also switch electric or electronic signals. They were used to be fabricated on a piece of silicon. The circuit was made by connecting these transistors together into a single silicon surface. The shape of circuits in an IC was printed together in all layers of silicon at the same time. This process takes the same amount of time even if the number of transistors in the circuit was increased. The cost of production of IC was decided by the size of silicon and not the number of transistors. This reduced the price of products due to which manufacturing and selling of IC increased and thus benefits and sales also. From the idea of connecting individual transistors to the collection of these transistors (Logic Gates) and finally, the collection of these Logic Gates used to get connected into a single integrated circuit (IC). Nowadays, a single IC can even integrate small computers onto it.

Gordon Moore, co-founder of Intel, in 1965, discovered that the number of transistors on a silicon microprocessor chip had made twice as much every year while the prices were reduced to half since their invention. This is known as Moore's Law. Moore's Law is considerable because it means that computers and their computing power get smaller and faster over time. Though this law is putting the brakes on now and consequently, the improvement in classical computers is not like before it used to be [2].

This leads to the idea of the smallest computer by reducing the size of the circuit up to the size of an atom. But then these circuits will not be able to act as a switch as electrons inside an atom can become invisible from one side of a barrier and appear on another side, i.e. they can exist in more than one place at the same time. This is due to the teleporting phenomena in quantum mechanics called "Quantum Tunneling". It shows that the size of the circuits of the classical computer after 5–7 nanometers has reached their limit. The representation and processing of these computers can be illustrated by the law of classical physics that gives us an only deterministic justification of the Universe. But it fails to forecast all noticeable phenomena occurring in nature and this led to the discovery of quantum mechanics, the biggest changeover in physics. Thus, there is a need for new computing other than current classical computing to put its state into some physical information rather than a circuit. Since the quantum phenomena are bringing up more constraints on the design of the computers. It changes the basic building blocks of a computer that not only expects new type of hardware creation but also a new design, software, and layers of abstraction to facilitate the designers to create and exploit these systems even if their complexities scale over time. The design of the hardware components has to be governed by quantum properties [3].

Quantum Computing is a new kind of computing based on Quantum mechanics that deals with the physical world that is probabilistic and unpredictable in nature. Quantum mechanics being a more general model of physics than classical mechanics give rise to a more general model of computing- quantum computing that has more potential to solve problems that cannot be solved by classical ones. To store and manipulate the information, they use their own quantum bits also called '*Qubits*' unlike other classical computers which are based on classical computing that uses binary bits 0 and 1 individually. The computers using such type of computing are known as '*Quantum Computers*'. In such small computers, circuits with transistors, logic gates, and Integrated Circuits are not possible. Hence, it uses the subatomic particles like atoms, electrons, photons, and ions as their bits along with their information of spins and states. They can be superposed and can give more combinations. Therefore, they can run in parallel using memory efficiently and hence is more powerful. Quantum computing is the only model that could disobey the Church-Turing thesis and thus quantum computers can perform exponentially faster than classical computers.

3. Need for quantum computers

Quantum computers can solve any computational problem that any classical computer can. According to the Church-Turing thesis, the converse is also true that classical computers can solve all the problems of quantum computers too. It means they provide no extra benefit over classical computers in terms of computability but there are some complex and impossible problems that cannot be solved by today's conventional computers in a practical amount of time. It needs more computational power. Quantum computers can solve such problems in reasonably and exponentially lower time complexities, also known as "**Quantum Supremacy**" [4].

Peter Shor in 1993 showed that Quantum computers can help to solve these problems considerably more efficiently like in seconds without getting overheated. He developed algorithms for factoring large numbers quickly. Since their calculations are based on the probability of an atom's state before it is actually known. These are having the potential to process data in an exponentially huge quantity. It also explains that a practical quantum computer could break the cryptographic secret codes. It can risk the security of encrypted data and communication. It can expose private and protected secret information. But the advantages of quantum computers are also kept in mind that is significantly more than its flaws. Hence, they are still needed and further research is going towards a brighter future.

4. Fundamentals of quantum computing

While designing the conventional computer, it was kept in mind that transistors' performance especially when getting smaller, will be affected by noise if any type of quantum phenomenon takes place. They tried to avoid quantum phenomena completely for their circuits. But the quantum computer adapts a different technique instead of using classical bits and even works on the quantum phenomenon itself. It uses quantum bits that are analogous to classical bits and have two quantum states where it can be either 0 or 1 except it follows some quantum properties where it can have both values simultaneously leading to a concept of superposed bits.

5. Where the concept of bits came from?

Transistors are the fundamental construction blocks for an IC which are connected through wires in a circuit. They conduct electric signals between devices. The communication between transistors within an IC takes place through electric signals. The behavior of the signals is analog in nature. Therefore, their values are real numbers that change smoothly between 0 and 1. These electric signals can also interact with the environment resulting in noise. Therefore, a little change from 0 to 0.1 due to temperature or vibrations from the environment can drastically change the system's behavior. There are two types of noise present in the environment. The first type of noise results from energy instabilities occurring suddenly within the object like temperature above absolute zero Kelvin. These are fundamental in nature. Other types of noise are the consequences of signal interactions. This type of noise could have corrected or designed. But neither of them got designed nor corrected or maybe left intentionally uncorrected at the hardware layer. They are systematic in nature [5].

To overcome these noises in analog circuits, the IC is built with transistors in such a way that it could work on digital signals (binary bits) instead of analog signals. These circuits are called 'Logic Gates'. They perceive the electric signals containing values of real numbers as a binary digit or 'bit' of either 0 (low voltage) or 1 (high voltage). Registers are another type of Gate which stores a bit or the number of bits present in an input value to process further. Gates can remove noise from a signal by limiting the set of values a signal can hold. Constructing IC using logic gates rather than transistors simplifies the designing by creating a powerful circuit that is not sensitive to design and fabrication issues and facilitates abstraction to designers so that they can focus only on gate functions (Boolean functions) rather than circuit issues. Boolean functions are defined by the rules of Boolean algebra. They can use an automated design tool for mapping the required logic gates. A standard library containing a set of tested logic gates is integrated into the silicon chip design with the help of their manufacturing technology. Negligible error rates can be achieved using digital logic and standard libraries. This helps in making the design robust. Also, the data is encoded by adding some redundant bits in the memory using an error correction code. This code is checked at regular intervals to detect the error. It also helps in other traits of design like testing and debugging.

Quantum Bit or Qubit is the fundamental unit of quantum information that represents subatomic particles such as atoms, electrons, etc. as a computer's memory while their control mechanisms work as a computer's processor. It can take the value of 0, 1, or both simultaneously. It is a million times more powerful than today's strongest supercomputers. Production and management of qubits are tremendous challenges in the field of engineering. They acquire both, digital as well as analog nature which gives the quantum computer their computational power. Their analog nature indicates that quantum gates have no noise limit and their digital nature provides a norm to recover from this serious weakness. Therefore, the approach of logic gates and abstractions created for classical computing is of no use in quantum computing. Quantum computing may adopt ideas only from classical computing. But this computing needs its own method to overcome the variations of processing and any type of noise. It also needs its own strategy to debug errors and handle defects in design.

Qubit has two quantum states similar to the classical binary states. The qubit can be in either state as well as in the superposed state of both states simultaneously. There is a representation of these quantum states also known as Dirac notation [6].

In this notation, the state label is kept between two symbols $|$ and $\rangle$. Therefore, states are written as $|0\rangle$ and $|1\rangle$ which are literally having analog values and both are participating to give any value between 0 and 1 given that sum of probability of occurrence of each state must be 1. Thus any quantum bit wave function can be expressed as a two-state linear combination each with its own complex coefficient i.e. $|w\rangle = x\,|0\rangle + y\,|1\rangle$ where x and y are coefficients of both the states. The probability of the state is directly proportional to the square of the magnitude of its coefficient. $|x|^2$ is the probability of identifying the qubit state 0 and $|y|^2$ is the probability of identifying the qubit state 1. These probabilities when summed up must give a total of 1 or say 100% mathematically, i.e. $|x|^2 + |y|^2 = 1$.

6. Properties of quantum computing

In quantum physics, the quantum object does not exist in an entirely determined state. It looks like a particle but behaves like a wave when not being observed. This dual nature of particles leads to interesting physical phenomena. The state of any quantum object is expressed as a sum of possible participating states or a wave-function. Such states are coherent due to the interference of all the participating states either in a constructive or a destructive manner. Observation of quantum objects when they interact with some larger physical system results in the extraction of information. Such observation of quantum objects is called quantum measurement. Measurement can also result in the loss of information by disrupting the quantum state. These are some of the properties of quantum objects. Quantum objects referred here are the qubits in the case of quantum computing. The progress of any quantum system is regulated by Schrodinger's equation that tells us about the change in the wave-function of the system due to the energy environment. This environment is the system Hamiltonian which is a mathematical description of energies experiencing from all forces felt by all components of the system. To control any quantum system, there is a need to control this environment by isolating the system from the forces of the universe that cannot be controlled easily and by assigning energy within this isolated area only. A system cannot be completely isolated. However, energy and information exchanges can be minimized. This interaction with the outside environment can lead to loss of coherence and can result in "Decoherence" [7].

The properties are the conceptual rules and mathematical manifestations that describe the behavior of the particles. Quantum computers use three fundamental properties of quantum mechanics to store, represent, and perform operations on data in such a way so that it can compute exponentially faster than any classical computer. The three properties are given as follows [8]:

- Superposition

Superposition in quantum mechanics states that any two quantum states can be summed up (superposed) resulting in another valid quantum state. It is a fundamental principle of quantum mechanics. Oppositely we can say that any quantum state is the sum of two or more than two other unique states.

Superposition in quantum computing refers to the ability of a quantum system where quantum particle or qubit can exist in two different positions or say, in multiple states at the same time. It provides high-speed parallel processing in an unbelievable way and is very different from their classical equivalents that have binary constraints. The quantum computer system holds the information that exists in two states simultaneously. Qubits are brought into a superposition by influencing them with the help of lasers so that it can simultaneously store 0 and 1 at the same time. In classical computing, if there are 2 bits, the total possible values after combining we get are 4, out of which only 1 value is possible at any instant. But on the other hand, if there are 2 qubits in the quantum computer. The total possible values after combination are 4 and all are possible at once. It looks like unthinkable because it is not like gravity that can be proved easily just by looking at the falling of an apple. The laws of classical physics fail here because superposition only exists in the territory of quantum particles.

For example, when solving a puzzle-like maze, a quantum particle can decide to take the various paths at the same time using superposition. This process matches the function of the parallel computer. Due to this property, the qubit is able to navigate the maze in exponentially less time than a classical bit

- Entanglement

Entanglement in quantum mechanics is a physical phenomenon where two or more quantum objects are inherently linked such that measurement of one rules the possible measurement of another. In other words, a pair or a group of particles interacts or share spatial locality such that the quantum state of each particle cannot be characterized independently of the other particle's state in the same group even when they are separated by a large distance.

Entanglement is one of the important properties of quantum computing. It refers to the strong correlation existing between two quantum particles (physical properties of systems) or qubits. Qubits are linked together in a perfect instantaneous connection, even if they are isolated at any large distances such as located at the opposite ends of the Universe. They are entangled or defined with reference to each other. The fact is that the state of one particle influences the state of the other. It creates strong communication between qubits. Once they got entangled, they will stay connected even after separated at any distance. In classical computers, if bits are doubled, computational power also gets doubled. But in the case of Entanglement, adding extra bits to a quantum computer can increase its computational power exponentially. Quantum computer uses this property in a sort of quantum daisy chain.

Some examples of entanglement can be seen in nature such as electrons separated from each other at some distance inside an electron cloud are massively entangled with one another. If one electron is at both the states of spin-up and spin-down with each state having a probability of ½, a similar case is with the other electron.

- Interference

The property of interference in quantum computers is similar to wave interference in classical physics. Wave interference happens when two waves interact with each other in the same medium. It forms a resultant wave with either their amplitudes added together when they are aligned in the same direction known as constructive interference or a resultant wave with their amplitudes canceled out when waves are in opposite direction known as destructive interference. The net wave can be bigger or smaller than the original wave depending on the type of interference. Since all subatomic particles along with light pose dual nature, i.e. particle and wave nature both. The quantum particle may experience interference. If each particle goes through both the slits (Young's double-slit experiment) simultaneously due to superposition, they can cross its own path interfering with the path direction. The idea of interference allows us to intentionally bias the content of the qubit towards the needed state. However, it can also result in a quantum computer to combine its various computations into one making it more error-prone [9].

7. The topography of quantum technology

The quantum phenomena are not limited to just quantum computing but they apply to other technologies also including quantum information science, quantum communication, and quantum metrology. The progresses of all these technologies are mutually dependent on each other and can control as well as transform the entire quantum system. They share the same theory of physics, common hardware and related methods [10].

Quantum Information Science seeks the methods of encoding the information in a quantum system. It includes statistics of quantum mechanics along with their limitations. It provides a core for all other applications such as quantum computing, communications, networking, sensing and metrology.

Quantum Communication and networking concentrates on the conversation or exchange of information by encoding it into a quantum system to facilitate communication between quantum computers. Quantum cryptography is the subset of quantum communication in which quantum properties help to design the secure communication system.

Quantum sensing and metrology is the study and development of quantum systems. The drastic sensitivity of such a system to environmental nuisances can be utilized in order to measure important physical properties (e.g. electric and magnetic fields, temperature, etc.) more accurately than classical systems. Quantum sensors are based on qubits and are carried out using the experimental quantum systems.

Quantum computing is the central focus of this research which exploits the quantum mechanical properties of superposition, entanglement and interference to enact computations. In common, a quantum computer is a physical system that comprises a collection of qubits that must be isolated from the environment for their quantum state to stay coherent until it performs the computation. These qubits are organized and manipulated in order to enforce an algorithm and to achieve a result with high probability from the measurement of its final state.

Difference between classical computers and quantum computers [11].

Comparison key	Classical computer	Quantum computer
Basis of computing	Large scale integrated multipurpose computer based on classical physics	High speed parallel computer based on quantum mechanics
Information storage	Bit based information storage using voltage/ charge	Quantum bit (qubit) based information storage using electron spin
Bit values	Bits having a value of either 0 or 1 and can have a single value at any instant	Qubits having a value of 0,1 or sometimes negative and can have both values at the same time
Number of possible states	The number of possible states is 2 which is either 0 or 1	The number of possible states is infinite since it can hold combinations of 0 or 1 along with some complex information
Output	Deterministic- (repetition of computation on the same input gives the same output)	Probabilistic- (repetition of computation on superposed states gives probabilistic answers)
Gates used for processing	Logic gates process the information sequentially, i.e. AND, OR, NOT, etc.	Quantum logic gates process the information parallel
Scope of possible solutions	Defined and limited answers due to the algorithm's design	probabilistic and multiple answers are considered due to superposition and entanglement properties
Operations	Operations use Boolean Algebra	Operations use linear algebra and are represented with unitary matrices.
Circuit implementation	Circuits implemented in macroscopic technologies (e.g. CMOS) that are fast and scalable	Circuits implemented in microscopic technologies (e.g. nuclear magnetic resonance) that are slow and delicate

8. The architecture of quantum computer

Architecture can be seen as a blueprint. The architecture of the quantum computer is a combination of classical and quantum parts and can be divided into 5 layers where each layer is represented as the functional part of the computer (**Figure 1**).

- **Application Layer**- It is not a part of a quantum computer. It is used for representing a user interface, the operating system for a quantum computer, coding environment, etc. that are needed for formulating suitable quantum algorithms. It is hardware-independent.

- **Classical Layer**- It optimizes and compiles the quantum algorithm into micro-instructions. It also processes quantum-state measurement returned back from hardware in the below layers and gives it to a classical algorithm to produce results.

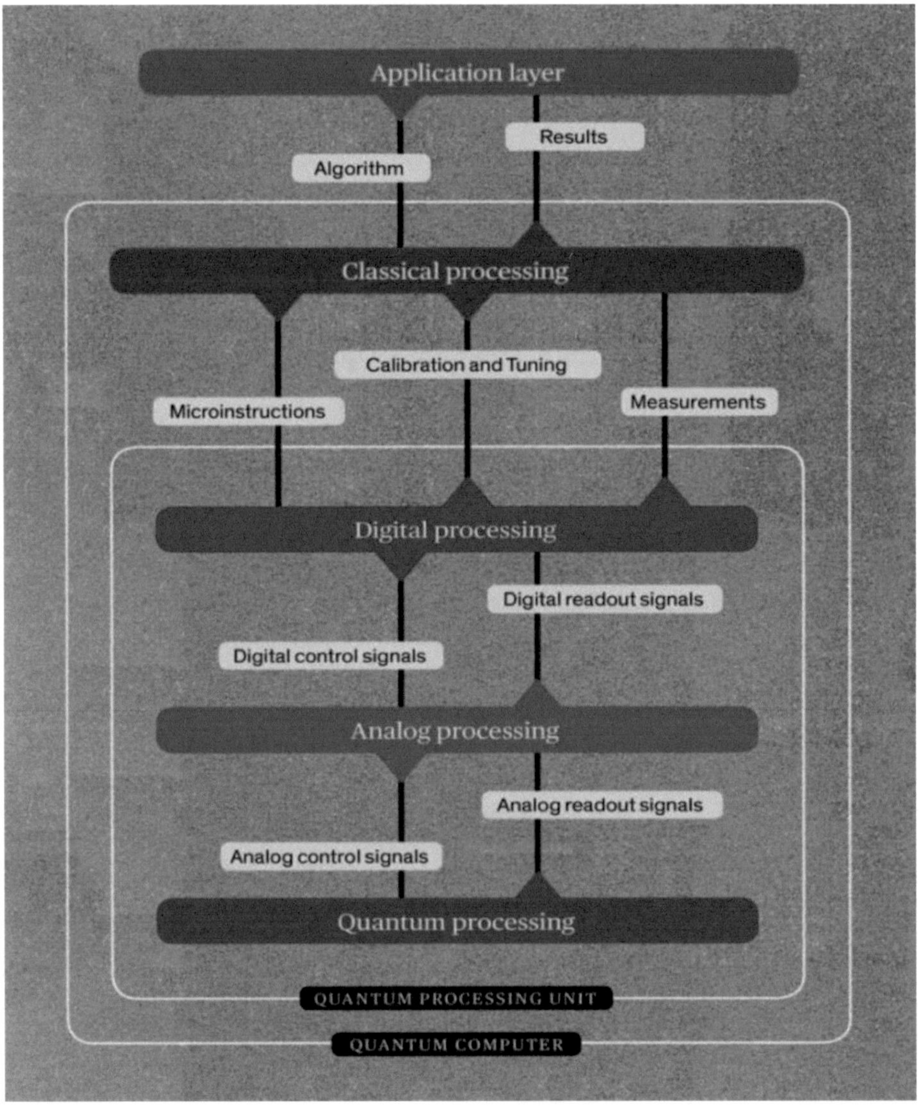

Figure 1.
The architecture of a practical quantum computer. It can be divided into five layers, each performing different types of processing [12].

- **Digital Layer**- It interprets microinstructions into signals (pulses) needed by qubit which act as quantum logic gates. It is the digital description of the required analog pulses in the below layers. It also gives quantum measurement as feedback to the above classical layer for merging the quantum outcomes to the final result.

- **Analog Layer**- It creates voltage signals which are having a phase and amplitude modulations like in wave, for sending it to the below layer so that qubit operations can be executed.

- **Quantum Layer**- It is integrated with the digital and the analog processing layer onto the same chip. It is used for holding qubits and is kept at room temperature (absolute). Error correction is handled here. This layer determines how well the computer performs.

Quantum Processing Unit (QPU) is made up of three layers including the digital processing layer, analog processing layer, and quantum processing layer. QPU and classical layer together constitute the Quantum Computer. Digital and Analog layers operate at room temperature.

9. Hardware and software of quantum computers

There should be an interface between the quantum computer and conventional computers for tasks related to data, networks, and users. In order to function usefully, the quantum qubit system needs organized control that can be managed by a conventional computer. The necessary hardware components for analog quantum computers are designed in 4 conceptual layers. First is the "quantum data plane" where qubit is present. Second is the "control and measurement plane" which is liable for performing operations and measurement on qubits as needed. The third is the "control processor plane" which defines the sequence of those operations and measurement outcomes to inform successive quantum operations required by the algorithm. And the last one is "host processor" which is a classical computer running a conventional operating system that handles user interfaces, network access, and big storage data structures. The processor is controlled using a high bandwidth connection that it provides [13].

A functional Quantum computer also requires software components in addition to the hardware. It is comparable to classical computers. Various new tools including programming languages are needed to substantiate quantum operations so that programmers can formulate algorithms, compilers that can map them to the hardware used by quantum computers and some other supports which can evaluate, optimize, debug and test programs. The programming language must be designed for any targeting quantum architecture. Some preparatory tools have been developed to support quantum computers and are accessible on the web [14]. These tools must be designed in an abstract way so that software developers can think more algorithmically without much concern for details of quantum mechanics. This software must be flexible enough to adapt to the changes in hardware and algorithms. This is one of the biggest challenges in quantum computing to develop complete software architecture. Other than programming languages, there must be simulation tools for modeling quantum operations and tracking quantum states and optimization tools for evaluating needed qubit resources so that it can perform different quantum algorithms in an efficient manner. The main goal is to minimize the number of qubits and the operations required for the hardware [15].

10. What is quantum algorithm?

An algorithm is a sequence of instructions or a set of rules to be followed to perform any task or calculation. It is a step-by-step process for solving a problem, especially by a computer. Any algorithm that can be executed on a quantum computer is called the **Quantum algorithm.** Generally, it is possible to execute all classical algorithms on quantum computers. However, the algorithms should contain at least one unique quantum step due to the property of either superposition or entanglement to be called a Quantum algorithm.

Quantum algorithms are characterized by a quantum circuit. A quantum circuit is a prototype for quantum computation that includes each step of the quantum algorithm as a quantum gate. A quantum gate is an operation that can be performed on any number of qubits. It changes the quantum state of the qubit. It can be divided into a single-qubit or multi-qubit gate, depending on the number of qubits on which it is applied at the same time. A quantum circuit is determined with qubit measurement [16].

An algorithm executing on a simulator rather than hardware is very profitable in terms of execution time by replacing the measurement overhead at the end of the algorithm. It is also known as simulation optimization. A quantum algorithm is always reversible when compared to the classical algorithm. It implies that if the measurement is not considered, a quantum circuit can be traversed back which can undo all the operations done by a forward traversing of the circuit. According to the undecidability problem, all problems that are unsolvable by a classical algorithm cannot be solved by quantum algorithms too. But these algorithms can solve problems significantly faster than classical algorithms. Some examples of the quantum algorithm are Shor's algorithm and Grover's algorithm. The Shor's algorithm can do factorization of very large numbers in exponentially faster than best-known classical algorithms [17], whereas, Grover's algorithm is used for searching large unordered list or unstructured databases that is four times faster than the classic algorithm [18].

There are various quantum algorithms available so far are as follows [19]:

- Fourier transform-based quantum algorithms
- Amplitude amplification-based quantum algorithms
- Quantum walks based algorithm
- BQP-complete problems
- Hybrid quantum/classical algorithms

11. Design limitations of quantum computer

The exponential computing power of quantum computers can be accomplished by assessing and rectifying any kind of design limitation which helps to avoid their quality degradation. There are four major design limitations. The first limitation is that the number of coefficients in Dirac notation that defines the state of a quantum computer rise exponentially with the rise in the number of qubits, only when all the qubits get entangled with each other. To obtain the full potential of quantum computing, qubits must follow the property of entanglement where the state of any qubit must be linked with states of other qubits. It cannot be achieved directly since it is hard to generate a direct relation between qubits. But it can be decomposed into a number of simple fundamental operations directly aided by the hardware. One can also perform indirect coupling which is known to be an overhead in machines in classical computing and is crucial at the early stages of development especially when qubits and gate operations are confined.

The second limitation is that it is impossible to copy an entire quantum system because of a principle called a no-cloning principle [20]. There is a risk of deletion of arbitrary information from the original qubits since the state of qubits or set of qubits are moved to another set of qubits rather than being copied. The generation and storage of copies of intermediate states or partial outcomes in memory is a necessary aspect of classical computing. But quantum computers need a different strategy. There are quantum algorithms that help to access classical bits from the storage so that it can be known which bits are loaded and being queried into the memory of the quantum system to perform its task successfully.

The third limitation is due to the absence of noise protection of qubit operations. The small deformities in gate operations or input signals are collected over time disturbing the state of the system because they are not discarded by the fundamental gate operations. This can highly affect the calculation preciseness, measurements and coherence of the quantum systems and lessen the qubit operations integrity [21].

The final limitation is the incapability of the quantum machine to identify its full state even after it has finished its operation. Assume quantum computer has introduced an initial set of qubits with the superposition of all states combination. After applying a function to this state, the new quantum state will have information about the function value for each possible input and measuring this quantum system will not give this information. Therefore, a successful quantum algorithm can be achieved by manipulating the system in such a way so that states after finishing the operations have a higher probability of getting measured than any other probable result.

12. Approaches to quantum computing

If we can design each gate slightly different from others, then the generated electric signals on communicating with each other produce periodic noise in each other. Thus, the noise immunity of gates used will be adequate to cancel the impact of various noise origins. Therefore, the concluding system will produce the same outcome as the logical gate model, even with millions of gates operating in parallel. The goal of the design is to minimize the noise in qubit that can prevent the qubit state to pass through noisy channels. The qubit state can be changed by changing its physical energy environment.

Thus, it leads to 2 approaches to quantum computing. In the first approach, the energy environment representing Hamiltonian is frequently changed smoothly as qubits operations are analog in nature and smoothly changes from 0 to 1 which cannot be completely corrected. It initializes the quantum state and then uses Hamiltonian directly to develop the quantum state. This is known as **'Analog Quantum Computing'.** It includes quantum annealing, quantum simulation and adiabatic quantum computers.

The second approach is similar to the classical computer approach where the problem is decomposed into a sequence of fundamental operations or gates. These gates have adequately defined digital outcomes for some input states. The set of fundamental operations of quantum computing is different from that of classical computing. This approach is referred to as **'Gate-based quantum computing'.**

13. Different categories of quantum computer

13.1 Analog quantum computer

This type of system performs its operation by manipulating the analog values in the Hamiltonian representation. It does not use quantum gates. It includes *quantum annealing, quantum simulation and adiabatic quantum computing*. The quantum annealing is done using some initial set of qubits that gradually changes the energy encountered by the system until the problem parameters are defined by Hamiltonian. This is done in order to get the highest probability final state of the qubits that corresponds to the solution of that problem. The adiabatic quantum computer performs computation using some initial set of qubits in the Hamiltonian ground state and then Hamiltonian is changed slowly enough such that it stays in its ground state or lowest possible energy while the process takes place. It has processing power similar to a gate-based computer but still cannot perform full error correction.

There are three basic types of analog quantum computing. These are divided on the basis of the required amount of processing power (number of qubits) and time to become practically and commercially available.

- Quantum Annealing

A basic rule of physics is that everything inclines towards a minimum energy state of a problem. This behavior is also true in the world of quantum physics. Quantum annealing is naturally used for real low-energy solutions such as optimization problems [22]. It is useful where the best solution is needed out of all possible solutions available. However, it is least powerful among all the types available. An example of this demonstrates an experiment to optimize traffic flows in a crowded city. Such an algorithm could successfully decrease traffic by choosing a convenient path. Volkswagen performs this with Google and D-wave system partnership. Such an experiment can be applied on a universal scale for all to get the cost-productive travel. This method can be applied to a collection of industry problems. For example, optimization of the flight route, petroleum price, weather and temperature information and passenger details, developing commercial aircraft.

Quantum annealing is also used for digital modeling, sampling problems and other science fields. This will take only a couple of hours to model all the individual atoms of air flowing over an airplane's wing at every tilts and speeds to formulate an optimized wing design. Using a sampling problem from energy-based distribution, the shape of energy can be characterized and is useful in machine learning problems. The samples improve the model using information about the state of the model for the given parameters.

- Quantum Simulation

Quantum simulations examine certain problems in quantum mechanics that are beyond classical physics. Simulating quantum phenomena that are complex in nature is one of the most important applications of quantum computing such as quantum chemistry. It includes modeling of chemical reactions on a large number of quantum subatomic particles. Quantum simulators can be used to simulate the misfolded protein structure [23]. Diseases like Alzheimer's are caused by misfolded proteins. Using random computer simulation, researchers test new treatment drugs and learn reactions. To achieve correctly folded protein structure and study all drug-induced effects, sequential sampling is done which could take more than a million years. Quantum computers can help evaluate it for making more effective treatments and medicines and it would be a significant healthcare improvement. In the future, quantum simulations will facilitate quick drug designing and testing by evaluating every possible drug combinations of protein.

- Adiabatic Quantum Computing

Adiabatic quantum computing is the most dominant, commonly applicable and hardest to create. A truly adiabatic quantum computer will use over a million of qubits. The maximum qubits we can access is less than 128 today. The basic idea behind this is that the machine can be directed at any complex calculation and obtain an immediate solution. This comprises analyzing the annealing equations, quantum phenomena simulation, etc. [24]. At least fifty unique algorithms other than Shor's and Grover's algorithm have been formulated to run on this quantum computer.

There is a possibility that quantum computers could revolutionize the area of artificial intelligence and machine learning. Some work has been done on algorithms that would operate as building blocks of machine learning but the hardware and software for quantum AI are still not practically accessible.

13.2 NISQ gate-based computer

NISQ stands for Noisy Intermediate-Scale Quantum. It is also known as the Digital NISQ computer. These type of systems are gate-based and operates on a collection of qubits without full error correction and cannot restrict all the errors. The computations must be designed in a way so that they remain practical on a quantum system with little noise and can be finished in fewer and sufficient steps so that Decoherence and gate errors do not hide the outcomes [25].

13.3 Gate-based quantum computer with full error correction

Such computers also perform gate-based operations on a set of qubits with the implementation of the Quantum Error Correction algorithm. It reduces or corrects the noise in the system occurring during the computation period. Errors may include inadequate signals, device forgery or undesired bonding of qubits to the environment or with each other. The error is reduced to such a limit that the system seems valid and precise for all computations. Such quantum computers can have various realizations and they must fulfill some conditions such as there must be an availability of a well-defined two-level system that can be used as qubits, a potential to initialize those qubits, a sufficiently extended amount of Decoherence time which can perform error correction and computation, quantum gates (a set of quantum operations) common for every quantum computation and a capability of measuring each quantum bit individually without bothering others [26]. The analog quantum computers and digital NISQ computers are in progress while the gate-based computers with full error corrections are much more difficult and demanding.

14. Advantages of quantum computing

1. According to researchers, quantum computers will be able to *solve those complex mathematical problems* that traditional computers find impossible to solve in a practical timeframe.

2. It provides that *computing power* which can sufficiently process excessively large amounts of data (2.5 Exabyte daily i.e. equal to 5 million laptops) created all around the world to extract meaning from it.

3. Due to the teleportation phenomenon known as 'quantum tunneling,' it can work in parallel and use less amount of electricity, hence, *reducing the power consumption* up to 100 to 1000 times.

4. A general quantum computer is "thousands of times" faster than any classical computer. For example, Google has made a quantum computer [27] that is 100 million times *faster* than any classical computer present in its lab.

5. It can solve complex problems *without being overheated* since for its stability it kept cold up to 0.2 Kelvin inside the quantum system.

6. It can easily *solve optimization problems* such as finding the best route and scheduling trains and flights. It would also be able to compute 1 trillion moves in chess per second. Quantum computers will be able to *crack the highest security* unbreakable encryption techniques. However, it would also build hack-proof alternates.

7. It can bring up *revolution* from drugs to petroleum industries. The invention of new drugs will become possible. The marketable algorithms of financial organizations can be improved. The field of artificial intelligence can be improved soon.

15. Disadvantages of quantum computing

1. Due to advancements in quantum computers, the *security* of the existing Internet of Things (IoT) would fall down. Cryptographic techniques, Databases of government and private large organizations, banks, and defense systems can be hacked. Considering these facts, quantum computers can be terrible for our future.

2. The Quantum Computer will work as a different device and cannot *replace classical computers* entirely. Since, classical computers are better at some chores than quantum computers like email, excel, etc.

3. It has *not been invented completely* yet as only parts are being implemented and people are still imagining how it would look.

4. It is very delicate and error-prone. Any kind of vibrations affects subatomic particles like atoms and electrons. Due to which noise, faults, and even failures are possible. It leads to "*Decoherence*" which is a loss of coherence in quantum.

5. Quantum processors are very unstable and are very hard to test even. For the stability of the quantum computer, it is kept at 0.2 Kelvin (absolute Kelvin) which is nearly below the universe temperature [28]. It is very hard to maintain and regulate such temperature. The main problem is to really develop it as a personal computer with the price range in the budget of consumers. They will be firstly accessible to large scale industry then come to retail markets.

16. Applications of quantum computing

Many quantum algorithms have been evolved for quantum computers that deliver speedup which is a result of some fundamental mathematical methods like Fourier transform, Hamiltonian simulation, etc. Most algorithms require a large number of qubits of the best quality and some error correction to provide useful functionalities. These algorithms are formed in blocks rather than as a whole combined application since it is not practical. Therefore, it is a great challenge to create quantum applications that are really practically useful along with providing speedup with no error. The potential utility or say useful application of a quantum computer is an area of ongoing research. It is predicted that those applications require fewer qubits and can be carried out with a lesser amount of codes. It is possible to build algorithms that can run faster on quantum computers because of the distinct features of the qubit. Below are some of the primary applications that we will see soon in the upcoming era:

- Cryptography

Many important elements of IT security and online security such as e-commerce and electronic secrecy depend on encryption and mathematical algorithms which are difficult to break such as factoring very huge numbers into primes (RSA technique). It is done by traversing through every possible factor using conventional computers which takes a significant amount of time. Also, some modern algorithms other than RSA like AES, ECDSA, etc. cannot be cracked using even high computing power. It makes it costly and cracking them even less practical. Quantum computers can do all these kinds of stuff in exponentially less amount of time. New quantum algorithms (e.g. Shor's algorithm) are able to do it and more unique algorithms will develop [29]. But before that, new encryption techniques are being made to resist the quantum ones. Since the already running techniques and digital applications security are at greater risks.

- Optimization Problems

Optimizing a problem implies finding the best solution to that problem out of all the possible solutions. It can be done by minimizing the error and even minimizing the steps available. Quantum computers are best in solving optimization problems. There are a lot of quantum algorithms out of which quantum optimization algorithms might improve the already existing optimization problems which are solved using conventional computers currently. Some of them are quantum semi-definite programming, quantum data fitting, and quantum combinatorial optimization. Some of the examples include simulating the molecular model like protein behavior for medical research which can lead to the new discovery of drugs for serious diseases like cancer, lung disease, etc. Another example is the Simulation of the cellular structure of batteries for improving battery power and life in electric vehicles. It could also solve travel-related problems in real traffic just like traveling salesman problems to find the shortest path between many cities, going to each city once and returning back, modeling the entire finance market, and many more. Traveling optimization is the major work under Volkswagen recently [30].

- Artificial Intelligence

Artificial Intelligence counts on processing large and complex datasets. It is responsible for learning, inferring, and understanding. It learns until it stops mistaking and making errors in its task. It takes a significant amount of time in learning too. But quantum computing can make it easy and more accurate. Since conventional computers are only training the learning model from a specific size of the dataset to restrict the computation time. Quantum computers can train these models over a huge dataset without sticking into the exponential time. The more data it uses to train, the more accurate it will be. Generative models generate output such as image, audio, etc. that can be fed to quantum computers to improve its quality and accuracy. Natural Language processing is another example that can understand complete sentences. Quantum computers can make it understand all the phrases and speech in real-time with improved quality, which is computationally costly with today's computer.

- Quantum Simulation

It is an important utility in the field of quantum chemistry and material science [31]. This problem needs solving ground state energies of electrons and their wave functions, with or without the presence of some external electric or magnetic field. From the structure of atoms and electrons in chemistry to the rate at which chemical reactions are taking place, everything can be simulated very well. The classical computer when applied to this problem often fails to reach the level of precision needed to predict the rate of the chemical reaction.

It could also have commercial applications in areas such as medical and healthcare fields, chemical catalysts, storage of energy, pharmaceutical advancement and device displays.

17. Major challenges in quantum computing

The good news is that at any instant of time, the quantum state with the same number of quantum bits can stretch over all possible states as compared to classical computers and thus works in an exponentially massive space. However, to be able to use this space requires all qubits to remain interconnected. Even after such progress, improvements are still needed. The bad news is that making new and high-quality qubits does not guarantee the creation and efficient use of fault-tolerant quantum computers and is still having challenges in its path [32].

Qubits cannot naturally ignore the noise. Hence, the quantum system is more error-prone. It suffers from *Decoherence*. The biggest challenge is how it can handle any undesirable deviations or noise in quantum computers. Classical computers can produce clean noise-free outcomes by simply putting its state as off or '0', which is not possible for quantum computers where errors occur in physical circuits. Qubits will gradually lose its information as well as interconnection (entanglement) between each other. The error rate is seen as a design parameter for such systems which should be improved in large qubit systems also. However, to make the qubits stable and error-free, they are being insulated from the outside environment in super-refrigerated fridges or vacuum chambers and accurately handled [33].

Qubits are neither completely binary nor digital. It is having analog properties also. Gate can reject noise by dealing with the input signal value of 0.8 and treating it as 1. But in the analog signal, every value between 0 and 1 is permitted since they have their meanings. Signals cannot be checked for any kind of noise or corruption. Since 0.8 can be 1 with some error or 0.8 without error. Presuming the error as 0 like Gates do or taking some noise value even if it was not present there can affect the adherence of the resulting quantum computation. Hence, there is a need for

DOI: http://dx.doi.org/10.5772/intechopen.94103

algorithms like quantum error correction similar to the logical error correction in classical computers. These algorithms can be run on a noisy gate-based quantum computer to eliminate the errors and noises present in them [34].

It is possible to employ a *Quantum Error Correction algorithm* on a quantum system. But quantum error correction requires dealing with the overhead such as a large number of qubits and their fundamental operations and generally needs more resources. Also, problems with large data inputs require a large amount of time to create the input quantum state that would monopolize the computation time lessening the quantum benefits.

Quantum algorithm development is another challenge since achieving quantum speedup expects entirely new types of algorithm design as the speed of computation depends on the design of the algorithm. The design of the algorithm should be corresponding to the number of qubits used.

Further *development of software tools* in addition to hardware, is required to create and debug quantum systems to help explain unknown issues and push towards designs.

Debugging quantum hardware and software is of utmost importance which depends on memory and intermediate machine states in classical computers. But in the case of quantum computing, states cannot be copied directly for later evaluation, and *directly measuring intermediate state can bring it to halt*. Hence, new strategies for debugging are essential for their development.

18. Importance of quantum computing

It is clearly possible to build a quantum computer that could perform computations that would run a lifetime on a classical computer. Practical applications of quantum computing need controlling the quantum phenomena and thus the quantum world to an exceptional level. This job requires substantial engineering and research to build, manage and employ a noiseless quantum system. The experiment with quantum supremacy is an important test of the theory of quantum mechanics that will help to improve the support of quantum theory and leads to unexpected discoveries. The development of aspects and components of quantum information technology and computing has already started to influence the area of physics. The quantum error correction theory to attain the fault-tolerant quantum system has proven important. The quantum information theory is practically useful to study physics and dynamics of multibody systems like a massive number of quantum subatomic particles and even in blackhole and related concepts. Advancement in this area is important for an accurate understanding of various physical structures. It has contributed to many other engineering fields like physics, mathematics, chemistry, computer science, material science, etc. It has also advanced classical computing. Strategies to develop a quantum computing algorithm have helped in improving the classical computing algorithm also. Research in the quantum algorithm has answered many questions in the computer science area. It can help to evaluate the safety of cryptographic systems, clarifying the limitations of physical computational and advancing computational methods. It will help to advance the human's understanding of the universe. The qubits that are recently being used in quantum computing is also used for building sensors, precision clocks, and other applications. Quantum communication is used for communicating two quantum systems at distance. There is an increased risk of asymmetric cryptography as well as the entire security system. Hence, the actions are being taken towards new quantum cryptography. The development of quantum information, science, technology and computing is a global area now.

19. Future scope of quantum computing

A significant amount of struggle is remaining before a practical quantum computer can be launched. There are some future advancements that are needed. Some of the future needs are enabling a Quantum Error Correction algorithm that requires low overhead and decreases the error rates in qubits, developing more algorithms with lesser qubits for solving problems, reducing circuit thickness so that NISQ computers can be operated, the advancement of methods which can verify, debug, and simulate the quantum computers, scaling the number of qubits per processor in such a way so that error rate is maintained or can be improved if possible, interleaving of operations in a qubit, recognizing more algorithms that can reduce the computation time and creating input–output for the quantum processor.

Such '*Quantum games*' are predicted in the future that will give unexpected situations and results that a player can experience because quantum computers will take all the possible operations and throws them into the game randomly due to its quantum properties like superpositioning and entanglement of qubits. It will be a never-ending experience.

'*Quantum computing in Cloud*' has the potential to overtake business initiatives like in other emerging technologies such as cryptography and artificial Intelligence. Since the classical simulation of fifty qubits is equal to the memory of one Petabyte that doubles with every single qubit added [35], the memory required should also be large enough to provide an environment for application development and testing for multiple developers to simulate quantum computers using suitable shared resources.

AI and machine learning problems could be solved in a practical amount of time that can be reduced from hundreds of thousands of years to seconds. Several quantum algorithms have been developed such as Grover's algorithm for searching and Shor's algorithm for factoring large numbers. More quantum algorithms are coming soon. Google has also declared that it would produce a workable quantum computer in the following 5 years with a 50-qubit quantum computer and will achieve quantum supremacy. IBM is also offering commercial quantum computers soon.

The progress of development in the field of quantum computers depends on many factors. Interest and financial support from the private sector can help developing commercial applications for NISQ computers. It depends on the progress of quantum algorithm development, availability of enough investment in the quantum technology field from government and the exchange of ideas within researchers, scientists and engineers [36]. To illuminate the limitations of quantum technology, a defensive result is also beneficial. It can help in overcoming those negative results which can lead to a new discovery.

Author details

Surya Teja Marella* and Hemanth Sai Kumar Parisa
University of Leicester, Leicester, UK

*Address all correspondence to: suryatejamarella@gmail.com

References

[1] Copeland, B. J. (2000). The modern history of computing, https://plato.stanford.edu/entries/computing-history/

[2] Theis, T. N., & Wong, H. S. P. (2017). The end of moore's law: A new beginning for information technology. *Computing in Science & Engineering*, *19*(2), 41-50.

[3] Richard P. Feynman, "Simulating physics with computers (1982)," International Journal of Theoretical Physics, Vol. 21, Nos. 6/7

[4] Arute, F., Arya, K., Babbush, R., Bacon, D., Bardin, J. C., Barends, R., ... & Burkett, B. (2019). Quantum supremacy using a programmable superconducting processor. *Nature*, *574*(7779), 505-510.

[5] Emily Grumbling and Mark Horowitz (2019),"2 Quantum Computing: A New Paradigm.", National Academies of Sciences, Engineering, and Medicine. Quantum Computing: Progress and Prospects. Washington, DC: The National Academies Press. doi: 10.17226/25196

[6] Charles H. Benett, and David P. DiVincenzo (March, 2000), "Quantum Information and computation," NATURE, Vo. 404, 16

[7] M.H.S. Amin, D.V. Averin, and J.A. Nesteroff, 2009, Decoherence in adiabatic quantum computation, Physical Review A 79(2):022107.

[8] Scott Amyx (2017), "quantum-computing-series-part-4-superposition-in-quantum-mechanics-381b98180f62", https://medium.com/@ScottAmyx/quantum-computing-series-part-4-superposition-in-quantum-mechanics-381b98180f62

[9] Margaret Rouse (2011), "Quantum Intereference", WhatIs.com, Tech Target https://whatis.techtarget.com/definition/quantum-interference

[10] J. Preskill, 2018, "Quantum Computing in the NISQ Era and Beyond," arXiv:1801.00862.

[11] Rajprasath Subramanian (2017), "10 Differences between Classical computing and Quantum computing," Medium, https://medium.com/@prasathbhuvana89/10-difference-between-classical-computing-and-quantum-computing-5e1777aa590d

[12] Versluis, Richard (2020, March), 'Here's a Blueprint for a Practical Quantum Computer', IEEE Spectrum, https://spectrum.ieee.org/computing/hardware/heres-a-blueprint-for-a-practical-quantum-computer

[13] Emily Grumbling and Mark Horowitz (2019), "5Essential Hardware Components of a Quantum Computer," Quantum Computing: Progress and Prospects, ISBN 978-0-309-47969-1 | DOI 10.17226/25196

[14] For example, QISKit and OpenQASM from IBM (https://www.qiskit.org/) and Forest from Rigetti (https://www.rigetti.com/forest)

[15] Emily Grumbling and Mark Horowitz (2019),"6Essential Software Components of a scalable Quantum Computer," Quantum Computing: Progress and Prospects, ISBN 978-0-309-47969-1 | DOI 10.17226/25196

[16] Mosca, M. (2008). "Quantum Algorithms". arXiv:0808.0369 [quant-ph].

[17] P. Shor, 1994, "Algorithms for Quantum Computation: Discrete Logarithms and Factoring," pp. 124-134 in 35th Annual Symposium on Foundations of Computer Science, 1994 Proceedings, https://ieeexplore.ieee.org.

[18] L.K. Grover, 1996, "A Fast Quantum Mechanical Algorithm for Database Search," pp. 212-219 in Proceedings of the Twenty-Eighth Annual ACM Symposium on Theory of Computing, https://dl.acm.org/proceedings.cfm

[19] "Quantum Algorithm," https://en.m.wikipedia.org/wiki/Quantum_algorithm

[20] W.K. Wootters and W.H. Zurek (1982), "A single quantum cannot be cloned", Nature 299(5886):802-803.

[21] T.P. Harty, D.T.C. Allcock, C.J. Ballance, L. Guidoni, H.A. Janacek, N.M. Linke, D.N. Stacey, and D.M. Lucas, 2014, High-fidelity preparation, gates, memory, and readout of a trapped-ion quantum bit, Physical Review Letters 113:220501.

[22] Fiona H (2018), "What is Quantum Annealing?", D wave Leap, https://support.dwavesys.com/hc/en-us/articles/360003680954-What-is-Quantum-Annealing-#:~:text=Quantum%20annealing%20is%20a%20heuristic,represent%20solutions%20to%20a%20problem

[23] Vineeth Veeramachaneni (2018), "Protein Folding: How Quantum Computing can help", Medium, https://medium.com/@veevinn/protein-folding-how-quantum-computing-can-help-6086b2456fb#:~:text=Protein%20folding%20is%20a%20problem,more%20quickly%2C%20and%20without%20limitations.

[24] A. Mizel, 2014, "Fault-Tolerant, Universal Adiabatic Quantum Computation," https:// arxiv.org/abs/1403.7694;

[25] J. Preskill, 2018, "Quantum Computing in the NISQ Era and Beyond," arXiv:1801.00862.

[26] D.P. DiVincenzo, 2000, The physical implementation of quantum computation, Fortschritte der Physik 48:771-783.

[27] David Nield (2015), "Google's Quantum Computer Is 100 Million Times Faster Than Your Laptop", Science Alert, https://www.sciencealert.com/google-s-quantum-computer-is-100-million-times-faster-than-your-laptop

[28] R. Barends, J. Kelly, A. Megrant, A. Veitia, D. Sank, E. Jeffrey, T.C. White, et al., 2014, Logic gates at the surface code threshold: Supercomputing qubits poised for faulttolerant quantum computing, Nature 508:500-503.

[29] Katwala, Amit (5 March 2020). "Quantum computers will change the world (if they work)". Wired UK.

[30] Vella, H. (2019). Quantum transforms travel. Engineering & Technology, 14(4), 50-53.

[31] Norton, Quinn (2007-02-15). "The Father of Quantum Computing". Wired.

[32] Franklin, Diana; Chong, Frederic T. (2004). "Challenges in Reliable Quantum Computing". Nano, Quantum and Molecular Computing. pp. 247-266. doi:10.1007/1-4020-8068-9_8. ISBN 1-4020-8067-0.

[33] M. Joseph, K. Elleithy and M. Mohamed, "A new Quantum Processor Architecture," 2019 IEEE 10th Annual Ubiquitous Computing, Electronics & Mobile Communication Conference (UEMCON), New York City, NY, USA, 2019, pp. 0483-0487. doi: 10.1109/UEMCON47517.2019.8992935

[34] A. Kandala, K. Temme, A.D. Corcoles, A. Mezzacapo, J.M. Chow, and J.M. Gambetta, 2018, "Extending the Computational Reach of a Noisy Superconducting Quantum Processor," arXiv:1805.04492.

[35] "Multiple Qubits (2017),"Microsoft Quantum docs, https://docs.microsoft.com/en-us/quantum/concepts/multiple-qubits

[36] Office of Science and Technology Policy, 2018, National Strategic Overview for Quantum Information Science, https://www.whitehouse.gov/wp-content/uploads/2018/09/ National-Strategic-Overview-for-Quantum-Information-Science.pdf.

Chapter 6

Multipoint-Interconnected Quantum Communication Networks

Qingcheng Zhu, Yazi Wang, Lu Lu, Yongli Zhao, Xiaosong Yu, Yuan Cao and Jie Zhang

Abstract

As quantum computers with sufficient computational power are becoming mature, the security of classical communication and cryptography may compromise, which is based on the mathematical complexity. Quantum communication technology is a promising solution to secure communication based on quantum mechanics. To meet the secure communication requirements of multiple users, multipoint-interconnected quantum communication networks are specified, including quantum key distribution networks and quantum teleportation networks. The enabling technologies for quantum communication are the important bases for multipoint-interconnected quantum communication networks. To achieve the better connection, resource utilization, and resilience of multipoint-interconnected quantum communication networks, the efficient network architecture and optimization methods are summarized, and open issues in quantum communication networks are discussed.

Keywords: multipoint-interconnected, quantum communication networks, quantum key distribution, quantum teleportation

1. Introduction

Quantum communication such as Quantum Key Distribution (QKD) and Quantum Teleportation (QT) is capable of exploiting the principles of quantum mechanics to transport classical, or even quantum, bits of information. quantum communication networks extend the concept of quantum communications, since they can transport, elaborate, and store quantum information (qubits) between different node pairs. Quantum communication networks leverage the principles of quantum mechanics including no-cloning, quantum measurement, entanglement, and teleporting. Hence, the new networking and computing capabilities emerge. At the same time, new and challenging constraints are imposed on the design and operations of quantum communication networks. This chapter firstly introduces the quantum communication enabling technologies including QKD and QT; then focuses on the research about QKD networks and QT networks to enable multipoint interconnection such as the architecture and service provisioning algorithms; finally, pays attention to problems and challenges of QT networking.

2. Quantum communication enabling technologies

The realizations of quantum communication network mainly include quantum key distribution technology and quantum teleportation technology.

2.1 Quantum key distribution

Quantum cryptography, which applies quantum properties to design the secure communication system, is the subset of quantum communication. QKD technology is a realization of quantum cryptography. It generates and distributes symmetrical cryptographic keys with information theoretical security based on the fundamental laws of quantum physics, i.e., the security is independent of all future advances of algorithm or computational power. QKD has the characteristic of "point-to-point" implementation. Thanks to the developments of quantum relay and switching technologies, the long-distance QKD is enabled. The following two subsections briefly introduce the QKD implementation and the related quantum relay and switching technologies to realize long-distance quantum communication.

2.1.1 Quantum key distribution implementation

The first quantum key distribution (QKD) protocol, the famous BB84 protocol, was proposed by Charles Bennett and Gilles Brassard in 1984 [1]. Since then, a series of QKD protocols such as E91, B92, SARG04, COW, DPS, GG02, MDI-QKD have been proposed one after another. There are three main implementation technologies of QKD: Discrete-Variable Quantum Key Distribution (DV-QKD), Continuous-Variable Quantum Key Distribution (CV-QKD), and Measurement Device-Independent Quantum Key Distribution (MDI-QKD). DV-QKD encodes information on a single photon and uses a single-photon detector for detection. DV-QKD originated earlier and is more mature, with a longer safe transmission distance. Besides, multi-node quantum network has been successfully established. The disadvantage is that single-photon sources are tricky to prepare [2]. Unlike DV or qubit-based QKD, the secret keys in CV-QKD are encoded in quadrature of the quantized electromagnetic field and decoded by coherent detections, which is lower cost and more practical. Under the same conditions, the output key rate of CV-QKD is much higher than that of the DV-QKD, and it is highly integrated with traditional optical communication networks. However, the current CV-QKD technology is not as good as the DV-QKD technology in terms of safe transmission distance, and the problem of working bandwidth also needs to be further resolved [3]. The security of MDI-QKD does not depend on whether the quantum device is trusted or not. MDI-QKD completely removes all security loopholes in the detection system and ensures a QKD network security with untrusted relays. Compared with CV-QKD, MDI-QKD can obtain higher security key rate, but the communication distance is shorter, and the channel is required to be asymmetric (that is, the measurement equipment is required to be close to the user on one side) [4].

In the past 10 years, a series of small-scale QKD technology verification networks have been built abroad, covering local area networks, metropolitan area networks, and intercity networks [5–9]. At the same time, a number of major technical research studies have been carried out in China to address quantum secure communication. Local area networks, metropolitan area networks, intercity networks, and wide area networks have carried out related work, including the quantum communication Beijing-Shanghai trunk line project for connecting metropolitan area networks, and the planned satellite-ground integrated wide-area quantum communication network. To keep faint quantum signals apart from intensive classical data signals, traditional QKD networks utilize low-noise dedicated fibers, such as dark fibers, which will significantly increase QKD deployment cost. Also, researchers have studied how to combine QKD deployment onto existing optical networks [10].

2.1.2 Quantum relay and switching

There are two main ways to achieve long-distance QKD, namely quantum relay technology and quantum switching technology. On the one hand, quantum relay technology can solve the problem of exponential attenuation of photon signal transmission in optical fiber for long-distance QKD. There are currently two types of quantum relay technologies. One is based on trusted relay, and the other is based on quantum relay. The trusted relay scheme is to cache the key generated by the point-to-point link in the trusted relay node and then transmit the end-to-end key required by the user hop-by-hop through the multi-hop link using one-time pad. This scheme breaks through the transmission distance limitation of the QKD link, but the relay node for key transmission must be trusted [11]. The quantum relay scheme is to use the principle of quantum entanglement to realize the storage and forwarding of quantum states, so as to realize the long-distance distribution of quantum states [12]. In order to overcome the fading of quantum information during quantum channel transmission, using quantum nodes instead of optical nodes to transform quantum information can effectively increase the transmission distance. Quantum nodes with this function are usually called quantum repeaters. This technology does not require trustworthy relay nodes, but it is still in the stage of theoretical research. On the other hand, in quantum switching technology, trusted relay is mainly used by switching nodes of quantum secure communication network based on single core fiber. Through relay nodes, the "Beijing-Shanghai trunk line" passes through Beijing, Jinan, Hefei, and Shanghai, connecting Beijing and Shanghai's quantum key distribution metro-network, which can provide data transmission based on quantum encryption for government affairs, finance, and other fields [13]. In 2018, Travis S. Humble et al. designed and implemented software-defined quantum networking protocol and soft switch to support the integration of quantum communication and existing optical communication [14]. In 2020, by integrating the fiber and free-space QKD links, the QKD network in China has been extended to a total distance of 4600 km, where any user in the network is able to communicate with any other [9].

2.2 Quantum teleportation

QT involves the transportation of an unknown quantum state from one location to another, without physical transfer of the information carrier [15]. It is one of the main technologies for constructing quantum communication networks.

2.2.1 Quantum teleportation implementation

QT is a quantum information transmission method using the uncertainty of quantum entanglement to realize the remote transmission of quantum states, which is one of the main technologies for constructing quantum communication network [15]. In 1993, Bennett et al. first proposed a theoretical protocol based on Einstein-Podolsky-Rosen entangled photons for teleportation [16]. The main idea is that the communication parties share a pair of entangled particles to establish a quantum channel, and the sender will transmit the unknown. After the quantum state and the shared particle perform a specific measurement on the local particle, the measurement result is notified to the receiving end, and the receiving end user performs a quantum gate operation on the particles owned based on the measurement result to obtain the quantum state to be transmitted by the sending end. It is worth noting that in the process of QT, the physical particles at the sender are not transmitted to the receiver but always stay in the sender. What is transmitted is only the quantum state, and the sender can even have nothing to do with this quantum state.

In 1997, the Zeilinger Research Group in Austria first reported the QT experiment in "Nature" [17]. The experimental results confirmed the feasibility of QT with a success rate of 25%. Since then, many scholars have developed theories of QT, exploring how to use different entangled states to construct quantum channels in the process of teleportation or how to transmit multi-qubit quantum states. In the current teleportation network experiment, the challenge is taken and a 30 km optical-fiber-based quantum network distributed over a 12.5 km area is constructed, which is robust against noise in real world with active stabilization strategies, allowing us to realize QT with all the ingredients simultaneously [18]. In Calgary fiber network, QT is reported from a telecom photon at 1532 nm wavelength, interacting with another telecom photon onto a photon at 795 nm wavelength. It improves the teleportation distance to 6.2 km [19].

2.2.2 Entanglement swapping and quantum repeaters

Quantum entanglement is a unique property of quantum systems, and it is also an important communication resource in QCNs. In principle, quantum entanglement is based on quantum superposition state [20]. Since quantum superposition experiments only reflect the indistinguishability of physical processes and are not limited to any specific physical quantities (such as momentum, energy, position, polarization, etc.), quantum entanglement is essentially not necessarily related to any specific physical quantities. The characteristics of quantum superposition have led many scholars to use a variety of methods to successfully prepare entangled states in experiments. For example, there are two typical methods for entangled photon generation technology based on parametric down conversion. The first type of entanglement source is the II-type phase-matched nonlinear crystal entanglement source [21]. The second entanglement source uses collinear nonlinear crystals to generate entanglement [22]. In addition, there is also the use of photonic crystal fibers to generate entangled photon pairs [23].

3. Quantum key distribution network

QKD generates and distributes symmetrical cryptographic keys with information theoretical security based on the fundamental laws of quantum physics, i.e., the security is independent of all future advances of algorithm or computational power. Quantum key distribution network (QKDN) is a network comprising two or more QKD nodes connected through QKD links, which allows sharing keys between the QKD nodes by key relay when they are not directly connected by a QKD link.

Although the international research on QKD is getting more and more in-depth, the research focus has always been on the performance improvement of the "point-to-point" QKD system, that is, how to increase the rate of quantum key generation, reduce the qubit error rate, and improve quantum key transmission distance, etc. It is difficult to use point-to-point QKD technology to support encryption requirements of various services from many nodes, and the security of services cannot be guaranteed. Therefore, it is urgent to establish a QKDN that supports multipoint interconnection. This part will introduce the existed research about QKDNs, including the QKDN architecture, trusted repeater node structure, routing and resource allocation, key pool construction, resilience, and machine learning application.

3.1 The QKDN architecture

Optical networks today represent a fundamental infrastructure for data transport in the Internet, with more than 2 billion km of fiber deployed globally. To integrate QKD into existing optical networks, an architecture of QKD-enabled optical network with software-defined networking technology is proposed [24], as shown in **Figure 1**. It satisfies the needs of key resource pooling, network openness, and pipeline flexibility. The architecture consists of four planes: application (app) plane, control plane, QKD plane, and data plane, in top-down order.

The application plane generates connection requests. It is at the top of this architecture and is the destination of the final application of quantum keys. It uses the shared key pair provided by QKDN to perform encrypted communication between users. It mainly includes two application types: key application and network application.

The control plane is implemented using an SDN controller and is in charge of resource management and allocation for the QKD plane and data plane. The control plane is the core module of the QKDN architecture. It controls the key distribution behavior of the QKD plane through the south-bound interfaces between the control plane and the QKD plane and communicates with the application layer. Introducing SDN is beneficial for managing the entire network's resources via logically centralized control. The north-bound interfaces of control plane open up network capabilities to the application plane. At the same time, the control plane can control the key supply strategies and complete the information interaction. Specifically, functions in the control plane of QKDN include QKDN topology acquisition, network virtualization, QKDN path calculation and resource allocation, QKD application registration, QKD service configuration, link control, policy control, notification processing, and quality of service control. The control plane also supports connection control, network optimization, and the ability to provide third-party applications in multi-domain, multi-technology, multi-level, and multi-vendor QKDN. In order to realize the scalability of the control plane, the control plane should also support hierarchical structure, multiple control domain division, and controller hierarchical nesting, etc.

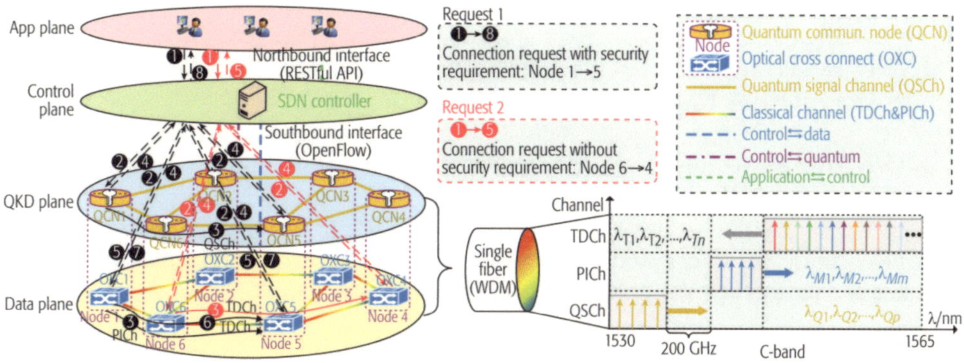

Figure 1.
The architecture of QKD-enabled optical network [24].

For each optical connection to be established in the network, in addition to the data channel (DCh), QKD requires a quantum channel (QCh) and a public channel (PCh) for secure key synchronization [25]. The QKD plane and data plane share fiber spectrum resources using WDM technology to construct QSCh, PICh, and TDCh. **Figure 1** shows a possible distribution of different channels in the fiber C-band. PICh and TDCh belong to the data plane. They can use general transmitters and receivers. Quantum communication node (QCN) has quantum switching functions: quantum signal sending and quantum signal receiving. It can use existing technologies for quantum switches, quantum transmitters, and quantum receivers [26]. Physically, an optical cross-connect (OXC) and a QCN are co-located at one node.

There are two types of connection requests in QKDNs including connection requests with and without security requirements. For example, when a connection request arrives with security requirements from node 1 to node 5 shown in **Figure 1** using black solid lines, SDN controller computes and allocates resources for channels including TDCh, PICh, and QSCh. In contrast, when the connection request arrives without security requirements from node 6 to node 4 shown in **Figure 1** using red solid lines, it is served by TDCh in data plane. The procedures of signals for configuring the two requests are delineated using black and red dashed lines in **Figure 1**, respectively. For the connection request with security requirement, the construction of QSCh and PICh for secure key synchronization is completed (steps 2–4), and the construction of TDCh is completed (steps 5–7).

3.2 Trusted repeater nodes structure

To overcome the distance limitation of QKD, either quantum repeaters or trust repeater nodes (TRNs) are required. However, the feasibility of quantum repeaters has yet to be demonstrated in practical long-distance QKD networks [27]. The TRN technique is a solution to construct long-distance QKD, and it has been widely adopted for the deployed QKD networks such as the deployed 2000 km QKD backbone network between Beijing and Shanghai in China recently.

An example of long-distance QKD based on TRNs is illustrated in **Figure 2** [28]. QBN_{src} and QBN_{dest} act as the source and destination QKD backbone nodes (QBNs) of two QKD users. TRN_1 and TRN_2 are deployed between QBN_{src} and QBN_{dest}. Three QKD links are separately established between QBN_{src} and TRN_1, TRN_1 and TRN_2, and TRN_2 and QBN_{dest}, while secret keys K_{s1}, K_{12}, and K_{2d} are separately produced on the three QKD links. To enable long-distance QKD between QBN_{src} and QBN_{dest}, four steps are performed as follows.

1. TRN_1 uses secret key K_{12} to encrypt secret key Ks1 and obtains the encrypted message $K_{12} \oplus K_{s1}$.

2. TRN_1 sends the encrypted message $K_{12} \oplus K_{s1}$ to TRN_2. TRN_2 uses secret key K_{12} to decrypt $K_{12} \oplus K_{s1}$ and obtains secret key K_{s1}.

3. TRN_2 uses secret key K_{2d} to encrypt secret key K_{s1} and obtains the encrypted message $K_{2d} \oplus K_{s1}$.

4. TRN_2 sends the encrypted message $K_{2d} \oplus K_{s1}$ to QBN_{dest}. QBN_{dest} uses secret key K_{2d} to decrypt $K_{2d} \oplus K_{s1}$ and obtains secret key K_{s1}.

Finally, QBN_{src} and QBN_{dest} can share the secret key K_{s1}. To guarantee the ITS of secret keys, one-time pad cryptosystem is required to be used for encryption. To

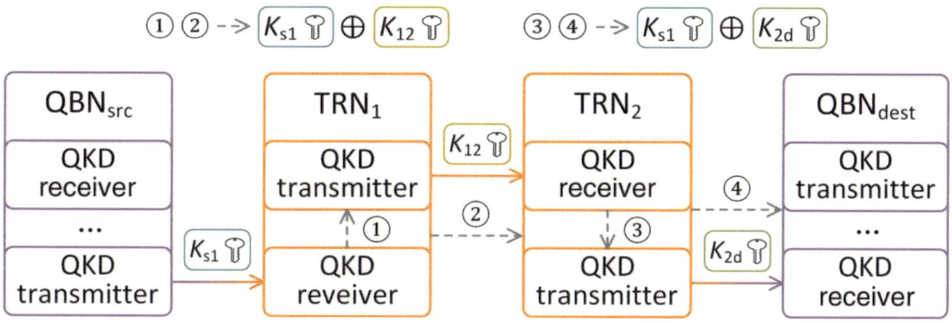

Figure 2.
Example of long-distance QKD based on TRNs [28].

extend the distance of QKD, a number of TRNs can be applied. Note that, each TRN is required to be trustworthy.

3.3 Routing and resource allocation in QKDN

With the expansion of the network scale, the number of users, and the continuous increase of security services, the problems of insecure key distribution process and low resource utilization in the key scheduling process in the prior art have become more and more prominent. To accomplish the key supply for services in QKDN effectively, the QKDN needs an efficient routing and resource allocation algorithm.

To accomplish the key supply for services, the concept of key as a service (KaaS) is proposed in [29]. Its meaning is providing secret keys as a service in a timely and accurate manner to satisfy the security requirements. The typical functions of KaaS are secret-key deployment and employment. To enable these functions, two secret-key virtualization steps are proposed including key pool (KP) assembly and virtual key pool (VKP) for secret-key deployment. For the KP assembly, the secret keys stored in each pair of key storages can be virtualized into a KP to facilitate secret-key resource management (e.g., KPA-B between KS-A and KS-B). For VKP assembly, a portion of secret keys in a KP can be virtualized into a VKP to enhance the security of dedicated service transmission (e.g., $VKP_{A\text{-}B\text{-}1}$ or $VKP_{A\text{-}B\text{-}2}$ abstracted from $KP_{A\text{-}B}$). Hence, with the combined two steps, the secret keys can be deployed and employed for securing different services in QKDNs.

Given that only finite wavelength resources can be reserved as QKD links, the time-scheduled technique can be applied to increase efficiency by dividing each wavelength channel for QCh/PCh into multiple time slots. Then, through the sharing of QCNs and QKD links in different time slots, the assembly of KPs can be realized between node pairs. The granularity of a time slot, which is denoted by t, is the synchronization time to produce a fixed number N of secret keys after KP assembly between two directly interconnected nodes. Note that, the synchronization time includes the time for channel estimation and calibration, qubit exchange, key sifting, and key distillation. Considering the constant consumption of the secret keys in KPs by the services for encrypting and decrypting data, the periodical KP assembly is needed to compensate for secret-key consumption. The period of KP assembly is denoted by T. Note that $t < T$, which ensures that KP assembly can be realized within a period.

As shown in **Figure 3**, a static time-shared KP assembly strategy for efficient secret-key deployment based on the Dijkstra and first fit (FF) algorithms is

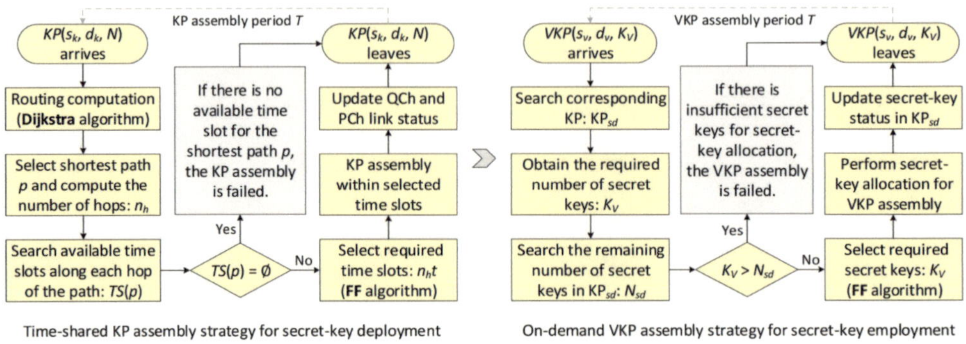

Figure 3.
KP and VKP assembly strategies for key supply [29].

presented. The KP assembly request is denoted by $KP(s_k, d_k, N)$, where s_k and d_k denote the source and destination nodes of the KP assembly request. The number of KPs is calculated by $n(n-1)/2$ since that KP is assembled between any pair of nodes. Here, n is the number of nodes in a QKDN. To compute and select the shortest QKD path between two nodes efficiently, the Dijkstra algorithm is utilized. The number n_h of hops is also computed, which aims to determine the required number of time slots. Then, to allocate available time slots for the assembly of different KPs, the FF algorithm is utilized.

After KP assembly, secret-key resource becomes a novel resource dimension in QKDNs, which can be virtualized. The virtualized KP is denoted as VKP. By assembling VKPs, the confidential services can be secured. Considering the different security requirements of services, different VKPs can require different numbers of secret keys. The type of VKPs with different secret-key resource requirements is denoted by V. The VKP assembly for secret-key employment is needed to satisfy the specific secret-key requirement of each VKP. The required secret keys for the assembly of a VKP are denoted by K_V. Secret keys will be updated and reallocated for VKP assembly when the KPs are assembled again. The updating and reallocating secret keys are necessary to enhance the security of confidential services.

Figure 3 presents a static on-demand VKP assembly strategy for efficient secret-key employment. The VKP assembly request is denoted by $VKP(s_v, d_v, K_V)$, where s_v and d_v are the source and destination nodes of the VKP assembly request. Secret-key resources cannot be reutilized, which are different from conventional computing, switching, and wavelength resources. Accordingly, some complicated resource allocation algorithms in conventional network scenarios such as most-used and load-balanced algorithms are not suitable for allocating secret keys in QKDNs. But the FF algorithm, which has high feasibility, can be utilized to allocate secret keys for the VKP assembly. The secret-key resources can be efficiently utilized and allocated using the on-demand VKP assembly strategy.

The simulations show the benefits of KaaS for efficiently deploying and employing secret keys as well as for security enhancement, where the balance of KPs' secret-key resources and VKPs' secret-key requirements can be achieved.

3.4 Key pool construction in QKDN

Aiming at the problem of low utilization of key resources in QKDNs, and the need to balance the inflow and outflow of key resources, a construction mechanism of virtual quantum key pools (QKPs) in QKDN is proposed [30], which achieves reasonable scheduling and efficient use of channel resources and key resources.

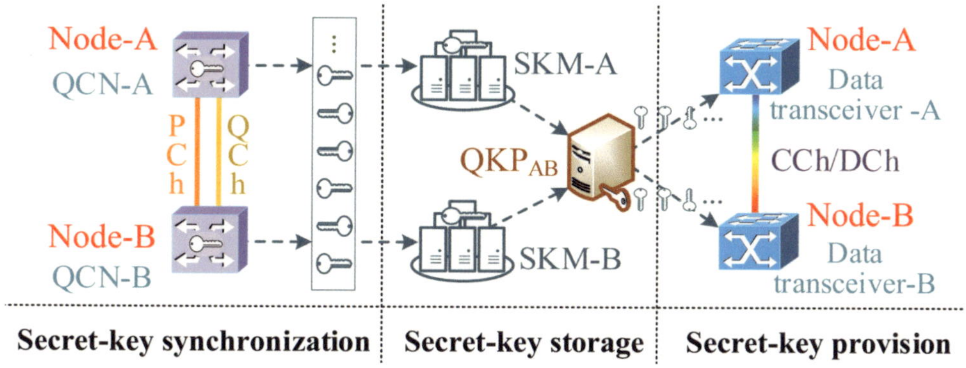

Figure 4.
QKP in point-to-point QKD system [30].

The extension of QKD from point-to-point systems to network-wide multipoint-interconnected systems requires to enhance the secret-key synchronization, storage, and provision, which improves the resource management and security performance. QKP construction in QKDNs is a potential solution to satisfy these requirements. In each node, there is a secret-key memory (SKM), which stores the synchronized secret keys. To improve the secret-key management, secret keys between each pair of SKMs are virtualized into a QKP, which is also denoted as VKP. QKP between the two nodes dynamically provides different numbers of secret keys for encrypting data according to different security requirements. **Figure 4** shows an example of QKP in point-to-point QKD system including QKD enhancements in secret-key synchronization, storage, and provision.

There are three main steps for constructing QKPs [30]:

- QCN-A encodes and transmits quantum signals to QCN-B via QCh.
- QCN-A and QCN-B interchange public information via PCh, so as to accomplish secret-key synchronization.
- After synchronization, SKM-A and SKM-B store secret keys between QCN-A and QCN-B respectively. The secret keys between SKM-A and SKM-B are virtualized to construct QKP$_{AB}$, which enables key supply on demand between Node-A and Node-B according to different security requirements through the CCh or DCh.

As for the support techniques for QKPs, QKPs are constructed on the control plane to manage the secret keys between QKD node pairs. They are all controlled by the SDN controller and can manage secret-key exchange, storage, assignment, and destruction. The SDN controller with programmable and flexible network control capabilities can also provide the effective implementation technique for QKPs.

4. Resilience of QKDN

The occurrence of failure is inevitable in QKDNs. Resilience of QKDN is very important. The key distribution on the corresponding routes will be disrupted, and key provisioning services will be affected by the failure of a single link. The security demands of users are intuitively violated. Apart from that, a high recovery time and capital expenditure will be indirectly induced further by such interruption.

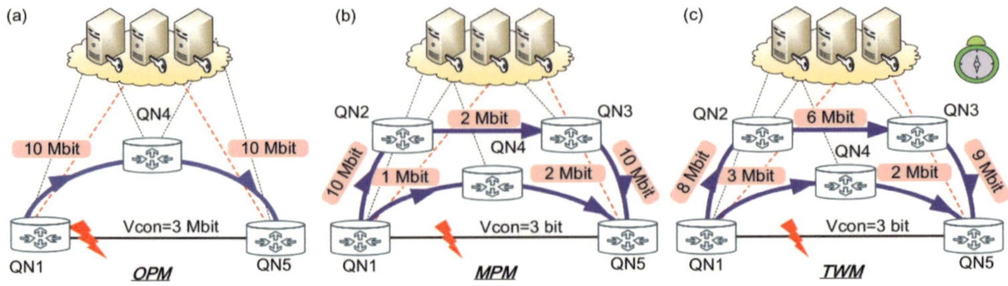

Figure 5.
Three methods. (a) OPM, (b) MPM, and (c) TWM [31].

Recovering and protecting failures for key provisioning services in QKDNs are an indispensable and vital problem to be solved.

In order to recover the key provisioning services affected by the failures in QKDNs, a Secret-Key Reallocation Strategy (SKRS) shown in **Figure 5(a)-(c)** is proposed including One-Path Method (OPM), Multi-Path Method (MPM) and Time-Window-based Method (TWM)) [31]. The strategy is to reallocate secret keys in QKPs and find available wavelengths, which are able to recover secret keys. By allocating the secret keys in QKPs over other paths, the security demand in failure-affected links will be satisfied. Multiple paths will try to provide keys simultaneously in case that the secret keys in one path are not enough. If multiple paths still fail to provide secret keys to meet the security demands, time division multiplexing technology can be considered. Simulation results verified that three proposed methods in the strategy can recover the failure-affected key provisioning services in different degree. Three methods of the strategy arc as follows.

- **One-Path Method (OPM)**: The secret-key provisioning capability of a path P is denoted by secret-key volume (K_{low}) provided in P. The two failure-affected nodes are taken as the source and destination. Multiple paths are calculated as set P for the recovery. For each p in P, K_{low} is calculated, and whether path P can satisfy the security demands (K_d) is checked. Here, the unqualified paths will be removed from P, and the rest of the paths will be sorted in the decreasing order of K_{low}. Then, the strategy tries to find wavelength resources that are able to recover, and it stops once the enough required resources are found, when the link between QN1 and QN5 fails, path QN1- > QN4- > QN5 will be chosen to provide the secret-key and wavelength resources, which is shown in **Figure 5(a)**. If no path is found, go to MPM.

- **Multi-Path Method (MPM)**: MPM uses multiple paths as a group to recover the failed key distribution services. Different from OPM, MPM needs to check whether the sum of K_{low} in set P can meet the security demands K_d. If so, OPM takes the paths that satisfy the K_d as the candidate recovery paths; otherwise, go to TWM. In case that no single path has secret-key provisioning capability, two paths QN1- > QN4- > QN5 and QN1- > QN2- > QN3- > QN5 jointly provide secret keys, which is shown in **Figure 5(b)**.

- **Time-Window-based Method (TWM)**: TWH retries the steps in the OPM and MPM since the volume of existing secret keys changes over time. The OPM and MPM are executed during the time window until they are successful.

4.1 Machine learning application in QKDN

Machine learning (ML) is an application of artificial intelligence (AI) that provides systems the ability to automatically learn and improve from experience without being explicitly programmed. In recent years, a huge amount of attention on ML has been attracted from both the academia and the industry. There has been much development related to ML technologies in both hardware and software. More on-board acceleration chips for neural networks are implemented by new low-power devices.

Due to the advantages of ML, ML can help to solve several problems in QKDN. In terms of parameter optimization for QKD, ML can greatly improve the efficiency of parameter optimization and allow it to be performed in real time on low-power devices, making it a highly useful tool for both free-space QKD and QKD networks. In terms of key resource utilization, the effectiveness of using reinforcement learning to realize resource allocation in QKD networks is verified, which was published by Asia Communications and Photonics Conference (ACP) 2020 and honored as the best paper award in industry innovation [32]. As for the standardization activities, *P* recommendation ITU-T Y.QKDN-qos-ml-req "Requirements of machine learning based QoS assurance for quantum key distribution networks" specifies the functional mechanisms of machine-learning-based quality of service (QoS) assurance for QKDN; the supplement ITU-T Y.supp.QKDN-mla "ITU-T Y.3800-series - Quantum key distribution networks - Applications of machine learning" specifies different application scenarios of ML in QKDN. In detail, the applications of ML in QKDN include the applications in the quantum layer, key management layer, and QKDN control and management layers of QKDN.

- **The applications of ML in the quantum layer of QKDN** represent applying the ML to improve the performance of the quantum layer such as the quantum channel performance: (1) ML-based quantum channel performance prediction method will predict the quantum channel performance according to different channel environments. Through the predictions, the quantum channel will be in the optimal performance state in real time. Measures can be taken in advance to improve the channel environment to reduce unnecessary losses. (2) ML-based QKD system parameter optimization solution will optimize the QKD system quickly and accurately based on the real-time changing environment, maintaining the QKD system in the optimal performance state in real time. (3) ML-based RUL prediction of components in a QKD system solution will accurately estimate the RUL of components, which greatly improves the operability of the components and provides a guarantee for the normal QKD system operations.

- **The applications of ML in the key management layer of QKDN** represent applying the ML in the key management layer and improving the key management efficiency and stability: (1) ML-based key formatting solution will reduce the time cost and the risk of key synchronization failure during the key consumption by guiding the key formatting with the awareness of service characteristics before storing keys. (2) ML-based key storage management solution will evaluate and predict health state of key storage and help to realize the efficient utilization of key resources. (3) ML-based suspicious behavior detection in the key management layer will improve the efficiency of suspicious behavior detection and achieve great authentication accuracy.

- **The applications of ML in the control and management layers of QKDN** represent applying the ML in the control and management layers and improving the QKDN management and control efficiency: (1) The ML-based data collection and data preprocessing will collect and preprocess multi-source, heterogeneous QKDN data in an efficient way. The collected and preprocessed data will be transformed into understandable, unified, and easy-to-use structures and optimized in the form of balanced characteristics for subsequent procedures. (2) ML-based routing solution will improve the routing effectiveness and the key resources utilization. (3) The ML-based QKDN fault diagnosis solution will reduce the loss and avoid the risk of QKDN faults by realizing fault location and fault prediction.

5. Quantum teleportation network

Quantum teleportation (QT) is a quantum information transmission method that uses the uncertain properties of quantum entanglement to realize the remote transmission of quantum states. This part will introduce the existing research about QTNs briefly, including the point-to-point QT mechanism and multi-Hop QT networking mechanism.

5.1 Point-to-point QT mechanism

Point-to-Point QT mechanism [33] is based on quantum entanglement exchange [34]. The basic principle of quantum entanglement exchange is as follows: The purpose of quantum entanglement exchange is to generate quantum entanglement between quantum systems that have never directly interacted through certain physical processes. Entanglement swapping has great application prospects in quantum communication and quantum information networks, such as preparing entanglement and extending the distance of quantum communication [35]. As **Figure 6** shows, suppose particles A, B and particles C, D are two sets of EPR entangled pairs, respectively. Performing the Bell basis joint measurement of particles B and C, particles A and D is also in an entangled state and is in the same entangled state

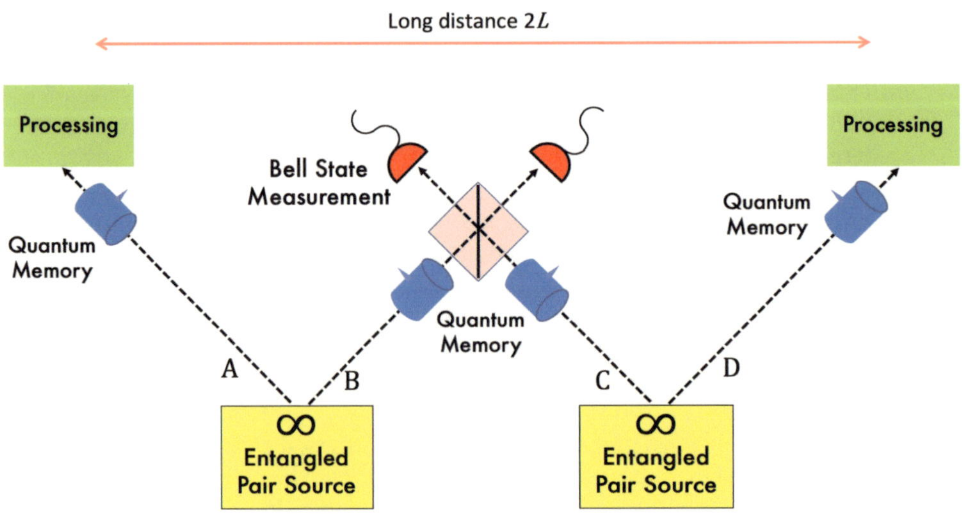

Figure 6.
Example of a one-hop, first-generation quantum repeater [36].

as particles 2 and 3, so that the entanglement exchange is successfully realized, and the point-to-point QT mechanism is completed.

5.2 Multi-hop QT networking mechanism

In a QTN based on teleportation, the necessary conditions for the transmission of information-carrying quantum states between two nodes are: a quantum channel composed of entangled particle pairs must exist between the source node and the destination node. However, it is impossible for any two nodes in the network to share entangled particle pairs, which means that the source node may not be able to directly transmit information to any other node in the network. In order to achieve communication between remote nodes, intermediate nodes are introduced to assist in the transmission of information [18]. Therefore, when the sender and the receiver directly share the entangled particle pair, the two nodes can directly transmit the quantum state; otherwise, there needs to be at least a quantum path established between the sender and the receiver—a quantum path established through an intermediate node. Entangled particle pairs are shared between neighboring nodes. The method of using teleportation technology to achieve quantum information transmission through intermediate nodes is called multi-hop QT [37].

In the traditional multi-hop QT system, the hop-by-hop QT scheme is often used. In the hop-by-hop QT transmission process, it is necessary to measure the entanglement of the nodes on the path one by one. According to the measurement result of the previous node, perform a unitary transformation on the particles held by the node to restore to the quantum state to be transmitted. The transmission of quantum information is from the source node to the destination node. An efficient multi-hop QT scheme is proposed [37]. The measurement results of the source node and the intermediate node are uniformly transmitted to the destination node, and only the unitary transformation is performed at the destination node.

6. Open issues in quantum communication networks

In recent years, the experimental research [15, 20, 38, 39] of quantum entanglement has mainly focused on how to expand the entanglement distribution distance, and the construction of quantum communication networks is mostly based on simple network topologies, mainly point-to-point network topologies, and a small number of star or bus network topology containing several nodes. There is little research on quantum communication networks under complex network structures, and research work should also be concentrated in the field of network security and quantum state transmission. Few works [40] have studied entanglement distribution from the level of quantum communication networks, so it is urgent and challenging to realize the connectivity of quantum communication network, repeating, switching, and routing quantum communication network and multi-layer quantum communication network.

- Connectivity of quantum communication network. How to deploy entangled particles and the location of distribution nodes play a vital role in the connectivity of the network. However, most of the current quantum communication network construction work is based on simple network topologies, mainly point-to-point network topologies, and a small number of star or bus network topologies that contain several nodes. There is little research on quantum communication networks under complex network structures, and research work is mainly focused on the field of network security and quantum state transmission, and the research on how to improve network connectivity is still insufficient.

- Repeating, switching, and routing quantum communication network. For the collaborative planning of the QT network, the main goal is to create relays, switches, and routes for quantum entanglement. The physical and software solutions in traditional networks are not suitable for quantum networks. The challenges they face include different forms of quantum entanglement generation and exchange, multi-user purification protocols, fusion and coordinated control, operation of traditional networks and quantum networks. To distribute entangled pairs between fixed target pairs, quantum repeaters need to be used to extend the distribution distance of entangled pairs. Unlike the operation of classical repeaters, quantum repeaters do not amplify photons in an entangled state during photon transmission. On the contrary, the quantum repeater can "jump" the entanglement property in the extra distance interval by consuming the resources of the second entangled pair. The innovation to achieve this is the quantum process of entanglement swapping.

- Multi-layer quantum communication network. The current quantum communication experiments rely on a set of devices with limited functionality and performance. However, to create wide-area and operational quantum networks, we need more capable devices with additional functionality. The devices are required to satisfy suitable requirements for reliability, scalability, and maintenance. Essential network devices to construct QTN include quantum memory, quantum switches, multiplexing technologies, transducers for quantum sources. Quantum memory should be improved with efficient optical interface and satellite-to-fiber connections; quantum switches should have high speed and low loss; transduction including microwaves is required, which is from optical and telecommunications regimes to quantum computer-relevant domain. Designing a quantum internet prototype capable of performing the aforementioned tasks will require developing a new quantum-updated version of the network stack.

7. Conclusions

In this chapter, the technologies to realize multipoint-interconnected quantum communication networks are summarized. Quantum communication enabling technologies including point-to-point QKD technologies and QT technologies are the basis to construct multipoint-interconnected quantum communication networks. As two typical quantum communication networks, quantum key distribution network (QKDN) and QT network are introduced respectively. In order to interconnect multiple points in QKDN, four sub-problems (i.e., architecture of QKDN, key supply in QKDN, resilience of QKDN, and machine learning in QKDN) are addressed. The architecture in QKDN consists of four planes: application (app) plane, control plane, QKD plane, and data plane, in top-down order. Key supply in QKDN needs a reasonable quantum key pooling mechanism and a balanced key resource scheduling strategy. Resilience of QKDN includes three methods to recover the failure-affected key provisioning services in different degrees. In order to interconnect multiple points in QT network, the existed research mainly pays attention on point-to-point QT mechanism and multi-hop QT networking mechanism, only few works have studied entanglement distribution from the multipoint interconnection of QT networks. Some open issues in quantum communication networks are also discussed, such as connectivity of quantum communication networks and how to plan quantum communication networks collaboratively.

Author details

Qingcheng Zhu, Yazi Wang, Lu Lu, Yongli Zhao*, Xiaosong Yu, Yuan Cao and Jie Zhang
State Key Laboratory of Information Photonics and Optical Communication, Beijing University of Posts and Telecommunications, Beijing, China

*Address all correspondence to: yonglizhao@bupt.edu.cn

References

[1] Bennett CH, Brassard G. Quantum cryptography: Public key distribution and coin tossing. In: IEEE International Conference on Computers, Systems and Signal Processing; 9-12 December 1984; Bangalore, India; Theoretical Computer Science. Vol. 560. 2014. pp. 175-179

[2] Yuan Z, Plews A, Takahashi R, Doi K, Tam W, Sharpe A, et al. 10-Mb/s quantum key distribution. Journal of Lightwave Technology. 2018;**36**(16): 3427-3433

[3] Tobias EA, Takuya H, Benjamin P, Georg R, Ruben L, Mikio F, et al. Wavelength division multiplexing of 194 continuous variable quantum key distribution channels. Journal of Lightwave Technology. 2020;**38**(8):2214-2218. DOI: 10.1109/JLT.2020.2970179

[4] Yin H, Chen T, Yu Z, Liu H, You L, Zhou Y, et al. Measurement-device-independent quantum key distribution over a 404 km optical fiber. Physical Review Letters. 2016;**117**(9):190501

[5] Chai G, Huang P, Cao Z, Zeng G. Suppressing excess noise for atmospheric continuous-variable quantum key distribution via adaptive optics approach. New Journal of Physics. 2020;**22**(10):103009

[6] Zhou X, Zhang C, Guo G, Wang Q. Improved decoy-state measurement-device-independent quantum key distribution with imperfect source encoding. IEEE Photonics Journal. 2019;**11**(3):7600207

[7] Peev M, Pacher C, Alléaume R, et al. The SECOQC quantum key distribution network in Vienna. New Journal of Physics. 2009;**11**:075001

[8] Aguado A, Lopez V, Lopez D, et al. The engineering of software-defined quantum key distribution networks. IEEE Communications Magazine. 2019;**57**(7):20-26

[9] Chen Y, Zhang Q, Chen T, et al. An integrated space-to-ground quantum communication network over 4,600 kilometres. Nature. 2021;**589**(7841): 214-219

[10] Cao Y, Yongli Zhao YW, Xiaosong Y, Zhang J. Time-scheduled quantum key distribution (QKD) over WDM networks. IEEE/OSA Journal of Lightwave Technology. 2018;**36**(16):3382-3395

[11] Piparo LN, Razavi M. Long-distance trust-free quantum key distribution. IEEE Journal of Selected Topics in Quantum Electronics. 2014;**21**(3):123-130

[12] Guo Y et al. Quantum relay schemes for continuous-variable quantum key distribution. Physical Review A. 2017; **95**(4):042326

[13] de Riedmatten H, Marcikic I, Tittel W, Zbinden H, Collins D, Gisin N. Long distance quantum teleportation in a quantum Relay configuration. Physical Review Letters. 2004;**92**(4):1-4

[14] Guo D, Liu X, Ma Y, Xiao L, Long G. A theoretical scheme for multi-user quantum key distribution with N Einstein-Podolsky-Rosen pairs on a passive optical network. Chinese Physics Letters. 2002;**19**(7):893-896

[15] Yonezawa H, Aoki T, Furusawa A. Demonstration of a quantum teleportation network for continuous variables. Nature. 2004;**431**:430-433

[16] Bennett CH et al. Teleporting an unknown quantum state via dual classical and Einstein-Podolsky-Rosen channels. Physical Review Letters. 1993;**70**(13):1895-1899

[17] Bouwmeester D et al. Experimental quantum teleportation. Nature. 1997;**390**:6660, 575-579

[18] Sun QC, Mao YL, Chen SJ, et al. Quantum teleportation with independent sources and prior entanglement distribution over a network. Nature Photonics. 2016;**10**(10):671-675

[19] Valivarthi R, Zhou Q, Aguilar GH, et al. Quantum teleportation across a metropolitan fibre network. Nature Photonics. 2016;**10**(10):676-680

[20] Pirandola S, Eisert J, Weedbrook C, et al. Advances in quantum teleportation. Nature Photon. 2015;**9**:641-652

[21] Kwiat PG et al. New high-intensity source of polarization-entangled photon pairs. Physical Review Letters. 1995; **75**(24):4337

[22] Fejer MM et al. Quasi-phase-matched second harmonic generation: Tuning and tolerances. IEEE Journal of Quantum Electronics. 1992;**28**(11): 2631-2654

[23] Fulconis J et al. Nonclassical interference and entanglement generation using a photonic crystal fiber pair photon source. Physical Review Letters. 2007;**99**(12):120501

[24] Zhao Y, Cao Y, Wang W, Wang H, Yu X, Zhang J, et al. Resource allocation in optical networks secured by quantum key distribution. IEEE Communications Magazine. 2018;**56**(8):130-137

[25] Lo H-K et al. Secure quantum key distribution. Nature Photonics. 2014;**8**:595-604

[26] Peev M et al. The SECOQC quantum key distribution network in Vienna. New Journal of Physics. 2009;**11**(7): 075001.1-075001.07500137

[27] Quantum Safe Cryptography and Security. ETSI White Paper No. 8, June 2015 [Online]. Available from: http://www.etsi.org/images/files/ETSIWhitePapers/QuantumSafeWhitepaper.pdf

[28] Cao Y, Zhao Y, Wang J, Yu X, Ma Z, Zhang J. Cost-efficient quantum key distribution (QKD) over WDM networks. IEEE/OSA Journal of Optical Communications and Networking. 2019;**11**(6):285-298

[29] Cao Y, Zhao Y, Wang J, Yu X, Ma Z, Zhang J. KaaS: Key as a service over quantum key distribution integrated optical networks. IEEE Communications Magazine. 2019;**57**(5):152-159

[30] Cao Y, Zhao Y, Colman-Meixner C, Yu X, Zhang J. Key on demand (KoD) for software-defined optical networks secured by quantum key distribution (QKD). Optics Express. 2017;**25**(22): 26453-26467

[31] Wang H, Zhao Y, Yu X, Chen B, Zhang J. Resilient Fiber-based Quantum Key Distribution (QKD) Networks with Secret-key Re-allocation Strategy. San Diego, CA, USA: OFC2019; 2019

[32] Zuo Y, Zhao Y, Yu X, Nag A, Zhang J. Reinforcement Learning-based Resource Allocation in Quantum Key Distribution Networks. Beijing, China: ACP/IPOC2020; 2020

[33] Huo M et al. Deterministic quantum teleportation through fiber channels. Science Advances. 2018;**4**(10):eaas9401

[34] Pirandola S. End-to-end capacities of a quantum communication network. Communications Physics. 2019;**2**(51)

[35] Jun Y. The Research on Quantum Teleportation of Quantum Communication. Huazhong University of Science and Technology; 2007

[36] Kleese van Dam K. From Long-distance Entanglement to Building a Nationwide Quantum Internet: Report

of the DOE Quantum Internet Blueprint Workshop. No. BNL-216179-2020-FORE. Upton, NY (United States): Brookhaven National Lab. (BNL); 2020

[37] Zhenzhen Z. Research on Multi-Hop Transmission and Networking for Quantum Communication Network. Southeast University; 2018

[38] Valivarthi R, Puigibert M, Zhou Q, et al. Quantum teleportation across a metropolitan fibre network. Nature Photon. 2016;**10**:676-680

[39] van Loock P, Braunstein SL. Multipartite entanglement for continuous variables: A quantum teleportation network. Physical Review Letters. 2000;**84**:3482

[40] Joshi SK et al. A trusted node–free eight-user metropolitan quantum communication network. Science Advances. 2020;**6**(36):eaba0959